Praise for

Agricultural Beginnings in the American Southwest

"This excellent book applies a range of conceptual approaches and an up-to-date data set to the analysis of agricultural origins—an important topic in the prehistory of the American Southwest and, more broadly, in world prehistory. I am not aware of any other recent, single-author publication that provides such a comprehensive review of the subject from across the American Southwest, including such thorough treatment of evidence scattered among what are often obscure (though generally well done) reports from the realm of contract archaeology."

—Richard Ahlstrom, HRA, Inc., Conservation Archaeology

"A standby for students of Southwestern archaeology at both the undergraduate and graduate levels. Roth's writing is clear, direct, and eminently readable."

—Francis E. Smiley, Northern Arizona University

"A much-needed and excellent review of one of the most significant processes in human history. Although studies of the Southwestern Archaic have generally been neglected, a variety of new insights have been suggested through current cultural resource management projects across the region. Most important, we are beginning to see regional variability in past foraging societies that encompasses the integration of cultigens into existing subsistence economies. Roth's new book brings together varied data sets and provides new perspectives to highlight regional variability and the diversity of forager-farmer behavior."

—Bradley J. Vierra, Statistical Research, Inc.

"Roth weaves a complex, variable, and diverse story of the absorption of agriculture into the Southwest. In addition to extensive discussion of hard data, both old and new, she examines several theoretical aspects that touch on critical issues. These include how (and why) agriculture was adopted by local hunters and gatherers, ritual and ideology, social issues, demography, why agriculture was more successful in some regions than others, methods of agricultural transmission from Mexico, and the often ignored but important concept of interactions between farmers and non-farmers. This will be a significant and insightful resource for years to come, not only to Southwestern archaeologists but to anyone interested in the complex story of how agriculture changed the world."

—Alan H. Simmons, Distinguished Professor of Anthropology, University of Nevada, Las Vegas

Issues in Southwest Archaeology

Edited by John Kantner

Issues in Southwest Archaeology features volumes that critically evaluate current archaeological research in the U.S. Southwest and Northwest Mexico. Titles investigate pervasive themes both in the archaeology of the region but also in contemporary anthropological inquiry, such as ethnicity, gender, migration, and violence. Authors discuss not only what archaeologists already know about the prehistory of the Southwest, but they also consider issues that impact the practice of archaeology today, including the roles of cultural resource management, oral history, and cultural property rights. Each contribution to the series is ultimately synthetic, comparative, and fully engaged in broader anthropological interests.

Editor John Kantner may be contacted at the following address:
Associate Vice President for Research & Dean of the Graduate School
University of North Florida
Office of Research 3/2501
Jacksonville, FL 32224

BOOKS IN THE SERIES:

Agricultural Beginnings in the American Southwest, by Barbara J. Roth (2016)

The Archaeology of Art in the American Southwest, by Marit K. Munson (2011)

Living Histories: Native Americans and Southwestern Archaeology, by Chip Colwell-Chanthaphonh (2010)

AGRICULTURAL BEGINNINGS IN THE AMERICAN SOUTHWEST

BARBARA J. ROTH

ROWMAN & LITTLEFIELD
Lanham • Boulder • New York • London

Published by Rowman & Littlefield
A wholly owned subsidiary of The Rowman & Littlefield Publishing Group, Inc.
4501 Forbes Boulevard, Suite 200, Lanham, Maryland 20706
www.rowman.com

Unit A, Whitacre Mews, 26-34 Stannary Street, London SE11 4AB, United Kingdom

British Library Cataloguing in Publication Information Available

Library of Congress Cataloging-in-Publication Data

Names: Roth, Barbara J., 1958– author.
Title: Agricultural beginnings in the American Southwest / Barbara J. Roth.
Description: Lanham, Maryland : Rowman & Littlefield, 2017. | Series: Issues in Southwest archaeology | Includes bibliographical references and index.
Identifiers: LCCN 2016029141 (print) | LCCN 2016042443 (ebook) | ISBN 9780759121713 (cloth : alk. paper) | ISBN 9780759121737 (Electronic)
Subjects: LCSH: Traditional farming—Southwestern States.
Classification: LCC GN407.4 .R67 2017 (print) | LCC GN407.4 (ebook) | DDC 306.3/49—dc23
LC record available at https://lccn.loc.gov/2016029141

∞™ The paper used in this publication meets the minimum requirements of American National Standard for Information Sciences—Permanence of Paper for Printed Library Materials, ANSI/NISO Z39.48-1992.

Printed in the United States of America

To my father, Ted Roth,
for his encouragement and stalwart support

CONTENTS

Figures

Tables

Acknowledgments

THIS BOOK IS THE PRODUCT of many years of thinking about, talking about, and studying early agriculture in the Southwest. As a result, many people have contributed to it, either directly or indirectly. First, thank you to John Kantner for asking me to do this book project and to John and Leanne Silverman for being so patient with me when it took forever to complete. I also want to thank Paul Fish, who got me interested in the Archaic period in the Southwest and provided me with opportunities to pursue research on it; John Speth, who taught me the importance of ethnographically informed archaeology; and Mike Schiffer, who taught me the importance of understanding context. Discussions with many colleagues and students over the years have helped shape my ideas about agriculture in the Southwest, and I owe special thanks to Bruce Huckell, Leon Lorentzen, Jonathan Mabry, Kevin Wellman, Dudley Gardner, Jenny Adams, Stephanie Whittlesey, and Nancy Coinman. Many people also shared their data and ideas for this book, for which I am ever grateful, including Alan Simmons, Brad Vierra, Chip Wills, Heidi Roberts, and Myles Miller. Aaron Woods and Ben Van Alstyne did the maps and figures and Roger Anyon, Dennis Gilpin, Myles Miller, Alan Simmons, Chris Ward, and Chip Wiles provided photos. I also want to thank the four anonymous reviewers whose helpful comments made this volume much stronger. Finally, I want to thank Rick, Sara, Lara, and my sisters Judie, Val, and Nanci for keeping me grounded and supported throughout this project.

Understanding Agricultural Beginnings in the American Southwest

1

THE ADVENT OF AGRICULTURE IS A TOPIC that continues to be of critical interest to archaeologists worldwide. Many volumes have addressed the important changes that occurred as agriculture was developed and the impacts that it had on prehistoric lifeways once cultivated plants became a component of the diet (e.g., Barker 2009; Kennett and Winterhalder 2006; Zeder 2006). One of the main reasons that this topic has been of such interest is the eventual impact that agricultural production had on later cultural developments in many parts of the world. Agriculture ultimately laid the foundation for fundamental changes in sedentism, population size, labor organization, and food security for populations who developed or adopted it. Understanding this critical transition in prehistory thus has broad implications for contextualizing the processes that led to later changes in economic, social, and political organization.

This fascination with the transition to agriculture is also apparent in studies in the southwestern United States. Agriculture was introduced to the Southwest after having been initially developed in Mexico, so the processes involved in its movement and the factors affecting the adoption of cultigens need to be examined separately from circumstances where agriculture was developed *in situ*. While not discounting the possibility that similar factors affected both the initial development and later diffusion of agricultural knowledge, Minnis (1992) and others have argued that crop acquisition was associated with its own unique set of circumstances. The Southwest provides an ideal setting for examining these circumstances, given the good preservation present at many sites and the long history of research.

The topic of agricultural beginnings in the Southwest has engendered considerable debate concerning the timing, routes, and the means of the introduction of agriculture. This book synthesizes data from sites across the Southwest with a focus on the diversity present in hunter-gatherer populations and environmental settings throughout the region and the influence that this diversity had on the adoption and integration of cultigens. Instead of addressing "the introduction of agriculture," I chose the title "agricultural beginnings" to emphasize this diversity. The theme of this book is thus variability—and encompasses variability in environments; in hunter-gatherer adaptations and cultural behaviors that preceded the arrival of cultigens; in the ways that cultigens were introduced; in reasons for adopting or rejecting cultigens; in the ways that cultigens were integrated into hunter-gatherer diets; and finally, in the impact of the use of cultigens on populations. By approaching the introduction of cultigens from this perspective, it is possible to evaluate the circumstances surrounding their initial use, integration, and impact—and examine situations where they were either ignored or had little initial impact. It then allows for exploring the consequences of the decision to incorporate agriculture in different portions of the Southwest.

The approach taken here is somewhat different from the approach taken in past studies of the introduction of agriculture in the Southwest as it focuses on the perspective of the hunter-gatherers and early farmers who integrated (or did not integrate) agriculture into their diet rather than on the process of the movement of agriculture into the region alone. Details on the initial development of maize agriculture in Mexico are not included, with the exception of the examination of possible zones of interaction.[1] This approach, as a longtime advocate of "peopling the past" (Tringham 1991), also allows for a more in-depth discussion of hunter-gatherer decision making than is possible by only examining the timing of the arrival and the changes in material culture that ensued. Models posited for the introduction of cultigens into the Southwest do not always take into account the people involved in this process and sometimes portray a lack of understanding of how hunter-gatherers behave. The goal of this book is to examine how hunter-gatherer and early farmer behavior can inform us about this important subsistence change, both in terms of the initial use of cultigens and the long-term effects of cultivation on these groups. This is done using evidence for the environmental and social settings in which agriculture was first used and the archaeological data associated with the earliest farmers in the region.

Early Agriculture in the Southwest

Domestication in the southwestern United States was all about plants. Domesticated dogs have been recovered from early farming sites in the Tucson Basin (Ezzo and Stiner 2000; Mabry 1998a), but they were not used as a food source. The only other domesticated animals used by southwestern peoples were turkeys, and these were not domesticated until after agriculture was introduced, although recent research in the northern Southwest indicates that they may have been domesticated by 100 BC (Speller et al. 2010). Animals were therefore not part of the earliest domestication process in the Southwest. This book thus focuses on the introduction of the earliest cultivated plants—maize (*Zea mays*), squash (*Cucurbita pepo*), and beans (*Phaseolus vulgaris*)—and discusses the possibility of the earlier manipulation and possible domestication of local indigenous plants such as amaranth (Minnis 1992, 2004).

Based on current data, maize was the first cultigen to appear in the Southwest. The presence of maize does not necessarily imply "agriculture," however, and the decoupling of the presence of cultigens from a subsistence regime based on agricultural production is necessary for examining the transition from foraging to farming in this region. The semantics of where to place early farmers has been a topic of concern to many researchers studying this transition in the Southwest. I follow Wills (1988a) in using the term *agriculture* in a general sense to designate the cultivation of domesticated plants, rather than distinguishing cultivation and horticulture from more intensive agriculture. This is in part because the intensity of production varied substantially across the Southwest after maize was first introduced. Over time, maize, beans, and squash, referred to by later puebloan groups as the "Three Sisters," became exceedingly important to many Southwestern peoples, and maize also became a fundamental component of ritual.

The region examined in this book is broader than recent overviews of the adoption of agriculture, which have generally focused on specific portions of the Southwest (e.g., Huckell 1995; Matson 1991), as this book synthesizes data on agricultural beginnings from the entire Southwest. Although hindered by the lack of available data on agricultural beginnings in some areas, much new archaeological data are available on the initial use of cultigens in this region, in large part because of cultural resource management projects. I have divided the geographic range covered in this volume both geographically and based on available data. The northern Southwest (figure 1.1) encompasses portions of the northern and southern Colorado

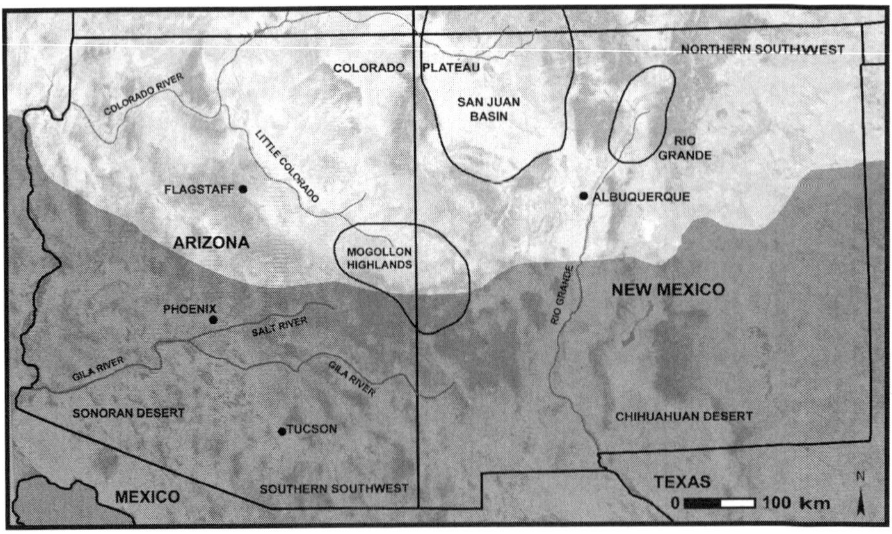

Figure 1.1. The northern and southern Southwest as defined in this volume. Figure by Aaron Woods

Plateaus in northeastern Arizona and southeastern Utah, the San Juan Basin of northern New Mexico and southwestern Colorado, the Albuquerque Basin, the northern Rio Grande, and the Mogollon Highlands. The southern Southwest (figure 1.1) encompasses the Sonoran and Chihuahuan Deserts, including the Phoenix and Tucson basins, southeastern Arizona, southern New Mexico, and portions of northern Mexico that have been included by other researchers as part of the Southwest, a topic that will be addressed in more detail in chapter 5. Because available data are irregular within these regions, the "strong" cases where most evidence is available are emphasized. These strong cases are from areas where archaeological research has revealed sites with evidence of early agriculture along with associated archaeological remains that can be used to evaluate the context of early cultivation. The emphasis on these strong cases is necessary, given the paucity of data from some areas such as northern Mexico and the Transition zone between northern and southern Arizona along the Lower Colorado River and Mogollon Rim, but it does not mean that these are the only areas where early agriculture occurred. Continued work in these understudied areas will likely expand and enhance our knowledge of the agricultural transition. As noted above, most of the recent information on agricultural beginnings comes from contract archaeology projects, and I have attempted to find, integrate, and synthesize these data as they are crucial to understanding this transition.

The chapters are organized to examine what pre-agricultural foragers were doing in each area, the context of the earliest agriculture, and the time period immediately following the initial introduction of agriculture. Agricultural beginnings are evaluated based on the environmental and cultural setting in each area where cultigens first appear and the circumstances that occurred after they were introduced; this allows for discussion of the differing reasons for the adoption of cultigens and differing responses to them.

Earliest Cultigens: Maize, Beans, and Squash

Current data indicate that the initial introduction of agriculture—which involved maize, and maize alone—was rapid and widespread (Smiley 1994), with multiple sites throughout the Southwest yielding radiocarbon dates on maize in the 2100 BC range (all dates presented here are calibrated BC dates unless otherwise specified). The impact of maize agriculture on hunter-gatherer populations was not consistent across the region, however, resulting in different levels of dependence and production intensity. In all areas, evidence for intensification is not associated with the initial arrival of maize, but instead occurs anywhere from five hundred to two thousand years later. Agricultural beginnings—the first maize—must therefore be decoupled from later responses that occur in both the areas where it was initially used and in areas where it spread.

The earliest maize was a popcorn variety with small cobs (figure 1.2), predominately of eight rows, and was characterized by low overall productivity (Adams 1994; Diehl 2005a; L. Huckell 2006). L. Huckell (2006) has documented few changes in early maize after its introduction and has remarked on the lack of variation present in early corn throughout broad regions of the Southwest.

Figure 1.2. Early maize cobs (adapted from photo by John Doebley). Figure by Russell Watters

Squash was domesticated in portions of Mesoamerica by 8000 BC (Ranere et al. 2009; Smith 1997), but its presence in the Southwest significantly postdates its initial use as a domesticate in Mexico. The earliest domesticated squash remains in the Southwest date to 1100 BC, based on direct dating of a squash rind found at

Sheep Camp shelter in the Chaco region of northern New Mexico (Simmons 1986) and a squash seed from Bat Cave in the Mogollon Highlands (Wills 1988a).[2] It is possible that the lack of earlier squash remains is tied to preservation, because of the way that squash was cooked and the lack of nonedible portions like maize cobs. However, given the widespread occurrence of maize in many disparate portions of the Southwest by 2100 BC, it is surprising that squash remains have not been recovered if, in fact, it was introduced at the same time as maize.

Beans appear even later in the Southwestern archaeological record with the earliest beans dating to 400–100 BC following their initial domestication in Mexico, which is currently dated to 1500–1300 BC (Smith 2001a).[3] The earliest beans in the Southwest have been recovered at Bat and Tularosa Caves in the Mogollon Highlands (Wills 1988a) and Fresnal Rock Shelter in southern New Mexico (Tagg 1996). This could be the result of preservation, as beans would be more likely to be preserved in protected dry caves and rock shelters, but it may also be tied to the route of introduction from Mesoamerica. DNA studies of maize indicate that it was first introduced via a highland route (Fonseca et al. 2015), and it is possible that the adoption of beans was the result of sustained interaction between early farmers across this route. Beans serve an important dietary role as they provide the essential amino acid lysine, which maize lacks, and thus are an important source of vegetable protein (Kaplan 1965). However, beans must be cooked, preferably by soaking and then boiling, to prevent stomach upset. Given the widespread occurrence of beans after ceramic cooking technology appears, it is likely that beans were not fully incorporated into the diet until the use of ceramic vessels made this possible. Both squash and beans require longer growing periods and more water than maize (Kantner 2004; Wills 1989) so they may not have been as attractive as maize was initially until better technology for water management and storage was developed.

If squash and beans were introduced hundreds of years later than maize, as is currently indicated, then the model of the introduction of a "crop complex" of maize, beans, squash, and bottle gourd first espoused by Ford (1981) and later Mabry (2005a) does not fit the initial introduction of cultigens to the Southwest. This does not mean that the crop complex was not important, as once the "Three Sisters" appear in the archaeological record, we see increasing levels of agricultural dependence and a more consistent focus on cultivation. However, as with many of the other factors discussed in this book, the initial use of each of these three cultigens must be evaluated separately. The later adoption of both squash and beans points to continued interaction with groups in Mexico after the initial adoption of maize. This kind of sustained interaction may have been important for later cultural changes as well, especially in the southern Southwest.

Models for the Introduction of Agriculture

Models for the introduction of agriculture to the Southwest have developed in sync with the recovery of early maize in different contexts. As will be discussed in more detail in chapter 2, the earliest model, developed following Dick's (1965) excavations at Bat Cave and Martin et al.'s (1952) excavations at Tularosa Cave in the Mogollon Highlands of west-central New Mexico, posited the introduction of maize to indigenous hunter-gatherer populations via a "highland corridor" that stretched from Mexico along the Sierra Madre Occidentale into the Mogollon Highlands, with subsequent movement to lowland settings after genetic changes resulted in increased drought tolerance (Haury 1962; Ford 1981). This model posited a very early entry of maize agriculture, as early as 3500 BC, with low levels of dependence and minimal impact on hunter-gatherer groups.

Subsequent challenges to the dating of this early maize by Berry (1982), additional work at Bat Cave leading to the dating of the earliest maize at 1000 BC by Wills (1988a), and data generated on early agriculture from excavations at large deeply buried sites along the major waterways in the Sonoran Desert (Huckell 1990, 1995; Mabry 1998a; Gregory 2001) led to a revision of this highland corridor model and essentially a split into two camps. One continued to view maize agriculture as introduced to indigenous hunter-gatherers with low levels of initial dependence, but posited a later 1000 BC entry (Minnis 1992; Wills 1988a,b, 1992). The second view posited that the changes observed in early agricultural sites along the Sonoran Desert waterways, including pithouses, large storage pits, and a plethora of new technologies, were so dramatic that they could not result from indigenous hunter-gatherers adopting agriculture, but instead must represent a migration of farmers from Mexico, with continued migration north onto the Colorado Plateau (Huckell 1995; Matson 1999, 2002). Linguistic data were later used to support this migration model (Hill 2001, 2002).

As additional data have come to light in the last decade, the timing of the introduction of maize has again been pushed back, now to 2100 BC, and the migration model no longer fits the data as the large agricultural sites along the Santa Cruz and San Pedro Rivers in the southern Southwest, first seen as coincident with the introduction of maize, now date between five hundred and eight hundred years later than the earliest sites with maize. The sites with the earliest evidence of maize are instead all seasonal hunter-gatherer occupations, more consistent with early experimentation with maize agriculture than with a full-scale agricultural adaptation. It is thus time to rethink our models of introduction and begin to examine the varying impacts that maize had on indigenous hunter-gatherer groups that led to the divergence in site characteristics observed across the Southwest by 1000 BC.

Basic Tenets for Examining Agricultural Beginnings

In the remainder of this chapter, I discuss several of the major ideas that have shaped our understanding of agricultural beginnings in the Southwest. The goal of this book is to both address and expand upon these basic tenets.

1. Foraging and Farming Behaviors

Agricultural beginnings must be understood in the context of Middle and Late Archaic foraging and farming behaviors. The Archaic period was defined by Willey and Phillips (1958) as a hunting-gathering stage prior to the advent of sedentary farmers. Much of what we know about the Archaic period in the Southwest today is based on the concept of the Archaic developed by early researchers working in the Great Basin and the Southwest. While this early work has been critical to our understanding of Archaic period occupations in different regions, it has also created some difficulties for researchers working on this time period, as many of the concepts associated with the Archaic period are rooted in the identification of "cultures" based on the description of lithic artifact assemblages, most often projectile points. Although more recent research has emphasized adaptation and variation, the roots of our knowledge of the Archaic, and especially the homogeneity in adaptation inferred for groups that share projectile point styles, have created a situation where adaptive diversity is somewhat muted in interpretations of Archaic period behavior.

Perhaps the greatest influence on interpretations of the Southwestern Archaic came from outside the Southwest. Jesse Jennings's excavations at Danger Cave in the Salt Lake Basin of east-central Utah led him to define the Desert Culture (Jennings 1957). The Desert Culture as conceptualized by Jennings represented a generalized hunting and gathering adaptation to an arid environment. Key components of this adaptation included high residential mobility, a focus on the exploitation of seed-bearing plants and small mammals, and the presence of small egalitarian bands. The technology used by these groups included small hand stones, basketry, bifacial knives, scrapers, bone tools, and projectile points. One of the critical aspects of the Desert Culture was stability in adaptation over a long temporal span, as Jennings (1957) argued that it was present by 9000 BC and continued into historic times in the Great Basin. The Desert Culture was viewed as representing an adaptation that characterized prehistoric hunter-gatherers living in arid environments throughout the West.

The Desert Culture concept has been criticized by researchers in both the Great Basin and Southwest (Heizer and Krieger 1956; Irwin-Williams 1967), but the idea that hunter-gatherers in this region were representative of a homogeneous adaptation to an arid setting has been pervasive and has structured the way many researchers think about the Archaic in these regions. This has occurred despite the fact that variability in the archaeological record has been consistently recognized and reported by researchers doing Archaic period research and the fact that a homogeneous Desert Culture adaptation contrasts significantly with what we know about forager behavior from ethnographic work with foragers (Kelly 2013).

Cynthia Irwin-Williams was the first researcher to openly challenge the use of the Desert Culture concept in the Southwest (Irwin-Williams 1967). She argued that by 3000 BC, distinctive regional adaptations were present across the Southwest that differed from the generalized Desert Culture concept. According to Irwin-Williams, these differences were based on hunter-gatherers adapting to local environments, which she saw as leading to cultural differentiation.

Irwin-Williams (1967) proposed the use of the term "Picosa" to classify Southwestern Archaic groups, taking the term from projectile point styles in the three major regions that she saw as distinct: Pinto Basin (west), Cochise (south), and San Jose (north). Although this term has not been used by later Archaic researchers, the idea that cultural differences existed among Archaic groups living in different environments throughout the Southwest has had an impact on the way researchers have approached the Archaic in the Southwest. The variability that is a key theme of this book is an attempt to sort out some of the differences between Southwestern Archaic groups and examine how these differences might have influenced the adoption of cultigens.

The locations of these earliest finds have also influenced our perception of agricultural beginnings. Most of the initial dates for early maize came from rock shelters, including Bat Cave in the Mogollon Highlands (Wills

1988a), Three Fir Rock Shelter on the Colorado Plateau (Smiley 1994), and Jemez Cave in the Rio Grande uplands of New Mexico (Ford 2013). This is in part a function of the level of effort expended on rock shelter sites, which were the focus of much early research in the Southwest because of the exceptional preservation of remains found in them, and in part a function of the preservation itself, which allowed for the discovery of maize cobs and kernels that were not always visible at open-air sites until the advent of flotation analysis. These rock shelters often contain a wealth of other data, including storage pits and cists, burials, and a large number of perishable remains including baskets, mats, rope, netting, and so on, that provide an excellent database for examining foraging and early farming behavior. One of the factors affecting our interpretations, however, has been the fact that these rock shelters are usually occupied as only part of a seasonal settlement and subsistence strategy (Wills 1995). Thus, the recovery of maize must be considered in light of what was going on during the entire year, rather than during the particular season or seasons when the rock shelter was occupied. This same problem is faced by researchers in the Tucson Basin, where early riverine agricultural settlements appear to have been part of a land use strategy that incorporated other resource zones (Roth 1992, 1995). Sites with evidence of early maize have now been found in multiple settings at both rock shelters and open-air sites and show the flexibility and adaptability of this new subsistence strategy.

2. *Environmental Circumstances*

Agricultural beginnings must be understood by considering the environmental circumstances surrounding the initial use of cultigens. The nature of the environment clearly impacted hunter-gatherer groups in the Southwest and consequently affected when, how, and to what degree cultigens were integrated into the existing diet. This is not to argue that the environment was the only factor influencing subsistence choices, but it is essential to recognize the crucial role that the environment played in providing both opportunities and constraints to prehistoric foragers and early farmers.

The Southwest is known as an arid region, but temperature, rainfall, and arable land are highly variable. The Mogollon Highlands, Rio Grande region, and San Juan Basin of the northern Southwest are characterized by mesas, canyons, and river valleys (figure 1.3). The challenges faced by hunter-gatherers and early farmers in these settings are different from those faced by groups to the south and included unpredictable rainfall, rocky and thin soils in some areas, and the potential for late frosts limiting early

Figure 1.3. Example of canyon and mesa environment in the northern Southwest—San Juan Basin. Photo courtesy of Dennis Gilpin

planting and early frosts killing harvests (Wills 1988a,b; Simmons 2006; Vierra 2008). Resource availability in this portion of the Southwest is highly seasonal and often spatially distinct, with key resources for hunter-gatherers including piñon (*Pinus sp*), juniper berries (*Juniperus sp*), walnuts (*Juglans sp*), acorns (*Quercus sp*), and large game such as deer (*Odocoileus sp*), antelope (*Antilocapra americana*), elk (*Cervus canadensis*), and mountain sheep (*Ovis canadensis*). The distribution of resources and the seasonal nature of their availability structured forager use of this environment and consequently impacted the adoption of agriculture in this region. Many of the earliest agricultural sites in the northern Southwest are found in the piñon-juniper zone or in alluvial settings near major washes (see chapter 4). It is likely that foragers occupied these zones to avoid scheduling conflicts between farming and foraging activities and to take advantage of the better soils and higher water tables in these settings (Wills 1992, 1995).

Further north in the mesas and canyons of northeastern Arizona and southeastern Utah, sustained occupation may not have been possible until agricultural productivity was high enough to permit it. Archaic period hunter-gatherers in this area focused on seasonally available resources such as piñon, Indian ricegrass (*Oryzopsis hymenoides*), and game, which were present in patches across the landscape, and the distribution of these resources and the timing of their availability resulted in relatively high levels of mobility (Geib 1996, 2011). The lack of reliable surface water made groups occupying this region more susceptible to drought, resulting in a reduction in hunter-gatherer use during dry periods. Agriculture in this setting would always have been risky, and this is perhaps why we see such a different pattern associated with the introduction of agriculture into this area (see chapter 4).

In the southern Southwest, desert environments prevail, and water was the critical factor influencing hunter-gatherer behavior and agricultural potential. The primary resources for hunter-gatherers in the Sonoran Desert of southwestern Arizona and northern Mexico include saguaro (*Carnegiea gigantia*) and other cacti (*Opuntia sp*), mesquite (*Prosopis sp*) and grasses along major waterways, deer (*Odocoileus sp*), cottontail and jackrabbits (*Sylvilagus, Lepus*) and a variety of small game (figure 1.4). Cienegas in alluvial settings may have been particularly attractive to foragers and early farmers because of the abundance of plants (cattails, bulrush, and sedges

Figure 1.4. Sonoran Desert environment. Photo courtesy of Pima County Office of Sustainability and Conservation

[*Typha sp*]), game, and the high water table (Hendrickson and Minckley 1985). In areas with abundant wild resources and closely spaced resource zones such as the Tucson Basin, foragers would have faced fewer scheduling conflicts than those in the northern Southwest, and this would have affected subsistence decisions and land-use strategies (Wills 1995). Roth and Freeman (2008) have argued that this resource structure, and especially the prolonged summer occupation of the floodplain, facilitated the integration of maize into forager diets in southern Arizona, as groups were occupying the floodplain to gather grasses and mesquite during the period when maize was tended and harvested, and they could easily incorporate maize into the existing subsistence regime.

In the Chihuahuan Desert to the east, desert grasslands are present with yucca (*Yucca sp*), agave (*Agave sp*), other cacti, mesquite, and grasses, along with small game and more widely dispersed large game (Doleman 2005; Miller and Kenmotsu 2004). Patchy but productive resource zones, including cienegas, were present near springs, rivers, and along playa shores, but these were often widely spaced across the landscape. The patchy resource distribution, dispersed nature of reliable water sources, highly seasonal nature of plant resource availability, and the general unpredictability of the environment tied to high aridity ultimately conditioned a much longer period of foraging activity in the Chihuahuan Desert than is seen elsewhere in the Southwest, even after the introduction of agriculture (Whalen 1994; Miller and Kenmotsu 2004). Other portions of the Southwest, such as western Arizona, did not witness the use of cultigens during the Archaic period at all, in large part because the arid environment and lack of reliable surface water precluded crop success.

Paleoenvironmental data point to variability in climate change throughout the Southwest. In most areas, it appears that the general vegetation pattern present today existed by 6000 BC (Van Devender 1990), with perturbations resulting from periods of drought interspersed with wetter periods. These droughts and wetter periods would have affected both foraging activities and farming potential, so fine-grained paleoenvironmental data are needed for each area to fully evaluate the interaction between environmental circumstances and human responses. Climatic data are available for some areas but not for others, although in some cases they can be extrapolated from existing data in nearby areas.

One of the biggest questions facing archaeologists interested in agricultural beginnings is the influence of the Altithermal on Archaic period foraging behavior. The Altithermal, or Middle Holocene drought, as defined by Antevs (1948, 1955), occurred from 5000 to 3000/2000 BC and, according to Antevs, resulted in the abandonment of large portions of the Desert West. Antevs's wide-sweeping statement of abandonment has been questioned by many researchers, who note that the drying period had differing effects in different portions of the West, including the Southwest. Wills (1995) argues that one of the most significant effects of climatic drying would have been the movement of the piñon-juniper zone from south to north, the expansion of grasslands, and the movement of cacti into the southern deserts as piñon moved north. He sees the response to Middle Holocene climatic change as resulting in regional differences in hunter-gatherer groups, but does not see it as necessarily detrimental to these groups. Huckell (1996a:319) has similarly argued that the use of the term *Altithermal* "at least oversimplifies and probably misrepresents middle Holocene climate." This contrasts with Mabry (1998b) and Carpenter et al. (2002), who argue that entire portions of the Southwest were abandoned in response to Altithermal drying, with groups moving to well-watered "refugia" in upland settings. These different interpretations come in part because Middle Archaic occupations are often dated using poorly dated projectile point styles with long temporal spans that may or may not represent Middle Holocene occupations.

The existing data argue against full-scale abandonment of the Southwest during the Middle Holocene. Instead, it appears that foragers responded to drying conditions in a variety of ways, including abandonment of some areas, reorganization of foraging efforts (Geib 1996; Meltzer 1995), and movement to reliable well-watered regions such as river valleys and upland springs. This may have led to a shift in subsistence strategies as well, with groups focusing more on plants than on large game, which may have been more heavily impacted by drying conditions, especially a reduction in the availability of surface water.

A number of researchers have noted that an increased focus on seed plants is a common response of foragers to drought conditions (Barlow 2002; Keeley 1995; O'Connell and Hawkes 1981), and this would have established a subsistence regime that may have been more conducive to the adoption of maize (Roth and Freeman 2008). Dello-Russo (2006) has proposed a Precipitation Threshold for Forager-Farmers (PTF), a quantified relationship between changes in precipitation and the productivity of wild seeds, that he argues can account for subsistence change during the Archaic period in some portions of the Southwest and may explain why some groups considered using cultigens. Drought-induced changes in the PTF would result in an increase in the use of wild seed plants, which Archaic period foragers could store as a component of an overwintering strategy. This restructuring of the subsistence base then resulted in land-use changes, including longer site use and increased logistical mobility involving task groups moving out from central base camps to hunt and gather. These same changes favored either the adoption of or intensified use of maize, especially in circumstances where precipitation changes were also associated with increasing population densities (Dello-Russo 2006:17).

Despite the debate over the influence of Middle Holocene drying on Archaic period adaptations, it is clear that the initial introduction of agriculture was associated with a wetter climatic period beginning between 2500 and 2000 BC across the Southwest (Mabry 1998b; Huckell 1996a; Miller and Kenmotsu 2004). In some areas this followed a long dry period, while in others it followed a period of climatic perturbations from dry to wet. The impact of this climatic uncertainty on foraging behavior was likely a central component in the decision to adopt cultigens. The wet period would have resulted in the expansion of resources that would facilitate the adoption of maize, including marshlands adjacent to farmable floodplains in the southern Southwest, grasslands in the uplands, and stabilized dunes throughout the region. During this period, springs would have recharged, and rivers and streams would have flowed more predictably. This wetter period would also have been associated with floodplain aggradation in many portions of the Southwest, making agriculture a more viable strategy (Dello-Russo 2006; Huckell 1996b).

Maize requires about 120 days for maturation and enough soil moisture to ensure seed germination (Wills 1988a:38). This was a challenge to early farmers in the Southwest, but the challenge differed depending on the environmental setting, with late frosts preventing early planting, and early frosts taking the harvest in higher elevation settings, and the lack of soil moisture affecting lowland settings.

The locations of sites with the earliest maize vary, which creates an interesting conundrum in trying to find patterns in the data, as it does not appear that the initial use of maize was associated with one particular environmental setting. The earliest maize in the Southwest currently comes from the Old Corn site (LA 137258) in west-central New Mexico, dating to 2100 BC (Huber 2005). The site is at an elevation of 6,200 feet in an open juniper woodland adjacent to a large wash. Other sites with maize dating to roughly the same time period are found in alluvial settings along the Santa Cruz River in the Sonoran Desert, a cave in the Gila Mountains of central Arizona, a rock shelter on the Colorado Plateau, an alluvial site in south-central New Mexico, and dune sites near Chaco Canyon.

The locations where early maize is present are primarily linked by the presence of some form of reliable water source, and this, above all other environmental factors, may have been the major influence on its initial adoption. Smith (1995:213) notes that early experimentation with agriculture occurred primarily in relatively rich environments, so the focus on wetlands and alluvial settings may have been part of this. This is supported by the fact that early maize has not been recovered from the most arid portions of the Southwest, such as southwestern Arizona. The other factor that these localities share is the presence of reliable seasonal resources, such as piñon in the uplands of the northern Southwest, chenopodium (*Chenopodium sp*) and amaranth (*Amaranthus sp*) in alluvial and dune settings, and mesquite and wetland resources in alluvial settings in the southern Southwest. These locations have implications for examining the factors influencing forager use of maize and the role of the environment in the transition from foraging to farming.

The broad-ranging similarity across the Southwest observed in studies of the earliest maize indicates that strains were not being developed for particular environments (L. Huckell 2006). Wills (1992:159) has also remarked on the fact that the genetic potential of maize was not exploited for the first one thousand years after its initial use. No segregation of seeds was made during the initial period of use, perhaps in response to the lack of reliable storage (Diehl and Waters 2006), or perhaps because it was not deemed necessary for a dietary supplement. However, Cummings et al. (2013) have identified several different varieties of early maize grown at the site of Las Capas in the Tucson Basin, suggesting that different varieties may have been grown in different settings to buffer crop failure.

L. Huckell (2006) sees gradual selection for an increased number of rows over a period of one thousand years after maize was first introduced. Diehl (2005a) has shown that cupule width increased slightly from the San Pedro (1200–800 BC) through Cienega (800 BC–AD 200) phases in the Tucson Basin; although these changes were minimal, they suggest some efforts toward increasing productivity by the end of the Late Archaic period. Increased selection and the eventual introduction of large-kernel "flour" maize around AD 700 had clear and important impacts on maize dependence (Diehl 2005a:370).

Wild resource use by Archaic foragers probably involved some level of plant manipulation, as has been observed ethnographically in Great Basin and Southwestern groups (Bean and Saubel 1972; Fowler 1996; Kelly 1977; Shipek 1989). Forms of plant manipulation noted in these groups included burning, broadcasting seeds, and transplanting. Some groups, such as the Cocopa along the Colorado River (Kelly 1977) and the Owens Valley Paiute (Steward 1933) practiced what has been termed "semi-cultivation" of grasses by scattering seeds over flooded areas. The Owens Valley Paiute added additional labor by constructing check dams and diversion ditches to flood a particular area, and then spread seeds over the area. No additional care was given to these plants between scattering and harvesting, so this activity cannot be classified as "agriculture" per se, but it represents the spectrum of activities associated with plants that we see represented archaeologically.

Several researchers have proposed that Archaic period foragers were doing similar kinds of plant manipulation prior to adopting cultigens, primarily with amaranth and chenopodium (Minnis 1992, 2004; Toll and Cully 1983; Wills 1995; Winter and Hogan 1986). Additionally, Wills (1995:27) notes that many Archaic period sites in the Mogollon Highlands and Colorado Plateau exhibit evidence of the consumption of a large quantity and variety of weedy annuals such as chenopodium, and argues that this is such a strong pattern that it may indicate that these were "encouraged or cultivated."

3. Social Context

The social context of agricultural beginnings must be considered in order to understand how agriculture began and flourished. All of the activities and changes discussed above took place within a social context that also significantly influenced the adoption of cultigens, the level of commitment to them, and the ensuing impact of cultivation. The very act of adopting implies the existence of social interaction, as hunter-gatherer groups in the Southwest were interacting with farmers in northern Mexico in order to obtain maize and the knowledge associated with growing it. This may have been accomplished in a number of different ways, including intermarriage, transmission via preexisting kin ties from previous interaction and movement, and exchange (Merrill et al. 2009; Roth 2006). Seasonal aggregation in resource-rich upland and alluvial settings may have provided social opportunities for the kinds of interaction that could result in the transmission of both maize itself and the techniques for growing it. Roth (2006) has discussed the important role that women played in this initial adoption, as their familiarity with plants, especially grasses that had characteristics similar to maize, and their processing and storing efforts would have influenced both whether maize was adopted in the first place and its ultimate successful incorporation into the diet.

The decision to cultivate necessitated certain labor requirements, which required either scheduling changes or the incorporation of maize into an existing resource procurement strategy. Wills (1992, 1995) has emphasized that maize was likely adopted most readily in areas where it did not require any dramatic changes in the resource schedule. The piñon-juniper zone and alluvial settings that were seasonally occupied and exploited with logistical mobility strategies, with groups occupying base camps and conducting gathering and hunting trips from these camps, would have been ideal candidates for this initial adoption (Roth and Freeman 2008). The decision to intensify production was likely more complicated, as even in areas with abundant seasonal resources, intensifying production via techniques like irrigation would have increased labor requirements and would have necessitated shifting labor efforts from some other subsistence tasks. The social context in which these resource decisions were made had a significant impact on how and when cultigens were adopted and integrated.

Summary: The Transition from Foraging to Farming

The goal of this book is not to answer all of the questions concerning agricultural beginnings in the Southwest but to synthesize what we know now and what that indicates about the process of the adoption and integration of agriculture. It is surprising that we still understand so little about agricultural beginnings, given the interest in the topic. This can be linked to several factors, including the emphasis on typology in Archaic studies, which has sometimes resulted in the interpretation of homogeneity across wide expanses of the Southwest; issues of preservation and site visibility, which impact our understanding of early agriculture due to both alluvial burial and subsequent occupations, as many later Ceramic period groups chose to live in the same localities as early farmers because of the agricultural potential; and issues of researcher interest, as the Archaic period has not generally drawn academic interest. The role of cultural resource management in our understanding of agricultural beginnings has been crucial, but has also sometimes been plagued by data accessibility, as these data are often presented in reports to sponsors that are not always widely distributed. Despite these problems, we now have a wealth of information on both Archaic period adaptations and early agriculture in the Southwest that, when synthesized, can provide substantial insights into the transition from foraging to farming.

These data have established that maize was first introduced to the Southwest by at least 2100 BC, yet three fundamental questions remain. The first is how maize first entered the Southwest. It is increasingly apparent that, initially, maize was introduced via adoption by various hunter-gatherer groups. The spread of agriculture then apparently occurred through a mixture of continued adoption by foraging groups and movement of farmers into new regions. The second question is why maize was initially adopted. In order to understand this, it is important to look at each portion of the Southwest separately, as the tremendous variability in environments and Archaic period adaptations resulted in differing responses to the availability of maize, with some groups adopting it quickly and others remaining hunter-gatherers until later in the sequence. The final question concerns the impact of agriculture and involves both efforts to intensify production and the addition of other cultigens, and this, too, varied across the Southwest. These issues are explored further in the remaining chapters.

Notes

1. Phytolith and plant DNA evidence from Mexico currently indicates that maize was initially domesticated from dense stands of teosinte in lowland tropical settings such as the Rio Balsas region of Guerrero by as early as 7000–8000 BC (Piperno et al. 2009; Ranere et al. 2009). Early squash remains have been documented at Xihuatoxtla Shelter dating to the same period (Ranere et al. 2009). These new data challenge previous interpretations that the earliest maize was developed in arid highland settings, based on the recovery of early maize at several well-known rock shelters in the Tehuacan Valley (MacNeish 1958) and Guila Naquitz (Piperno and Flannery 2001). The data indicate that maize was domesticated first in areas where dense stands of teosinte occur and later spread to highland settings. Squash remains recovered at Coxcatlan Cave dating to 5900 BC indicate that squash may have been widespread in its initial distribution (Smith 1997, 2005). The late arrival of squash into the Southwest is thus interesting, but it is possible that it is related to preservation factors.

2. Although squash remains are reported to have been found in deposits dating to 2000 BC at McEuen Cave (Merrill et al. 2009), no published data are available on this find and the remains were not directly dated, so it is not included here.

3. Half a possible bean cotyledon was found during excavations at the Las Capas site in the Tucson Basin dating to 1100 BC (Diehl 2005b); because of the lack of positive identification and the significant discrepancy in dating in relation to other finds, it is not included here.

How We Think about Agricultural Origins

2

OUR UNDERSTANDING OF AGRICULTURAL BEGINNINGS in the Southwest has been historically influenced by a focus on the timing of the introduction, with the result of sometimes minimizing the significance of other factors in this important transition. Despite Cynthia Irwin-Williams's (1967, 1973) call for examining local adaptations during the Archaic period, her ideas were not initially heeded by researchers interested in the introduction of agriculture to the Southwest, as much of research until the 1980s addressed the timing of the arrival of cultigens and the routes of transmission.

The Timing of Early Agriculture in the Southwest

The emphasis on the timing of the introduction of agriculture extends back to the era of early radiocarbon dating, when dated remains from two sites, Bat Cave (Dick 1965) and Cienega Creek (Haury 1957; Martin and Schoenwetter 1960) pushed back the arrival of maize to 4000–3500 BC during the Chiricahua, or Middle Archaic, period. Bat Cave, located in the Llano Estacado (Plains of San Augustin) of New Mexico (figure 2.1), was first excavated by Herbert Dick in the 1950s. Dick excavated in arbitrary 12-inch (50-centimeter) levels, and, as Wills (1988a) later noted, this resulted in substantial mixing of cave deposits. Maize found in the lower levels of the cave deposits was argued by Dick to be "primitive." Charcoal from the same levels, which was picked from the screen, was submitted for radiocarbon dating and returned early dates of 4000–3500 BC (uncalibrated). Middle Archaic period Chiricahua and Late Archaic period

Figure 2.1. Bat Cave. Photo courtesy of Chip Wills

San Pedro points were mixed in these levels, so they were assigned to the Chiricahua phase because of the dates (Wills 1988a).

The recovery of maize that dated to 4000 BC associated with an Archaic-period lifestyle of mobile hunting and gathering led to the notion that agriculture was 1) introduced from Mexico early to upland settings above 6,000 feet and this "primitive" maize was adapted to high elevations and moist environments, and 2) maize was a supplement to the diet of Archaic groups and did not become an integral component of the diet until the Ceramic period.

A second site used to illustrate that this early entry of maize was consistent across the Southwest was the Cienega Creek site, located in an alluvial setting in the Mogollon Highlands of central Arizona (Haury 1957). The site contained several hearths and forty-seven cremations, all of which were preceramic in age. One cremation yielded a radiocarbon date of 200 BC (uncalibrated). Below the cremations was a level containing a hearth, chipped stone, a Chiricahua projectile point, and several pieces of ground stone. The presence of the Chiricahua point was used to assign this component to the Middle Archaic period. Pollen samples from the ground stone yielded forty-two grains of maize pollen, again leading to the assumption that maize was introduced during the Chiricahua period (Martin and Schoenwetter 1960). Charcoal recovered from the Chiricahua levels was submitted to two separate radiocarbon labs, and they returned substantially different dates, with one dating to 500 BC and the second to 2200 BC (uncalibrated). Although the site was originally touted as evidence of early agriculture in the Southwest, the context of the remains, including the pollen washes from the ground stone, their questionable association with the dated charcoal, and the wide discrepancy in the radiocarbon dates all call this interpretation into question. Given the extent of the Late Archaic (San Pedro phase) occupation of the site, it is much more tenable that the maize remains date to the Late Archaic period, a contention supported by two of the three radiocarbon dates.

Challenges to the Early Entry Dates

Berry (1982) was the first to criticize the evidence used to document the early introduction of maize and to critically evaluate how maize might have been introduced to the Southwest. He found major discrepancies in reported radiocarbon dates from purported early agricultural sites, and his resulting reassessment argued that maize was not present in the Southwest until 500 BC and was not introduced onto the Colorado Plateau until 200 BC. His study was based on a small sample of early maize dates, however, and the recovery of numerous sites with directly dated early maize has filled some of the chronological gaps and refuted the interpretation of a late arrival of maize.

In the 1980s, Wills returned to Bat Cave, in large part because dates coming from work being done in the Southwest were not supporting the early date for the introduction of maize agriculture to the Mogollon Highlands from Dick's work there (Berry 1982; Ford 1981). The goals of Wills's excavations were to clarify the occupation sequence at Bat Cave and the archaeological relationships between the recovered projectile points, maize, and dates of occupation (Wills 1988a). He directly dated both maize and squash remains found in cave deposits using then newly developed techniques for directly dating cultigens. The maize returned dates of 1000 BC while the squash dated to 950 BC (uncalibrated; 1100–1200 calibrated BC), then the oldest squash remains in the Southwest. Wills (1988a:225) argued that there was a change in site use associated with the introduction of maize, with a shift from a more temporary seasonal occupation to a base camp after 1000 BC, visible archaeologically in the presence of storage pits and occupation surfaces.

By the 1980s, no sites with evidence of early agriculture that dated earlier than 1000 BC were known, and it appeared that maize was introduced relatively quickly to the Southwest and spread throughout the region after this time. This notion did not last long, however.

In the early 1980s, Simmons (1986) found maize at four small Middle Archaic sites in the San Juan Basin of northern New Mexico east of Chaco Canyon. Radiocarbon dates from two sites with features that contained maize pollen dated to 1700 BC (uncalibrated), suggesting a very early date for maize in this area. These dates were not initially accepted as valid, in part because they did not meet the new view that agriculture arrived at 1000 BC and in part because the dates were on charcoal from hearths so the association of the maize pollen with the charcoal was questioned. Charcoal from bell-shaped storage pits containing carbonized maize at another site dated to 770 BC, and direct dates on maize from Sheep Camp Rock Shelter came back at 200 BC, which were more in line with the dates on maize from other regional sites. What set these data apart was not only the hint of very early maize but also the fact that they were found at seasonal camps with no evidence of occupational intensity. Maize in this context was a dietary supplement; Simmons (1986) argued that it was planted in the summer in open grasslands and harvested in the fall before groups moved to winter in rock shelters.

In the 1990s, Smiley (1994) pulled together chronometric data from across the Southwest to examine the dating of early agriculture. He proposed that the initial spread of maize, then dating to circa 1000 BC, was likely "relatively rapid across great distances," but he also remarked on the lack of dates from the far northern Colorado Plateau (Smiley 1994:167). He proposed a sequence of occupation for the northern Southwest that began with the use of rock shelters by early farmers, establishing a long tradition that continued into Basketmaker times. This was followed by later Basketmaker movement into open-air sites. This dichotomy is perhaps most valid for specific portions of the northern Colorado Plateau. Recent dates on early maize from elsewhere in the Southwest have been obtained from open-air sites and thus, as with all things associated with agricultural beginnings, the situation appears to be more complicated than initially thought. Clearly rock shelter use is a major component of the agricultural transition, but it does not appear to be the exclusive domain of early farmers in the Southwest.

Routes of Transmission

A second and related research concern was the route (or routes) of transmission of early agriculture. Initially, Haury (1962) and later Ford (1981) posited a "highland corridor" for the entry of maize into the Southwest, with maize first introduced from northern Mexico via movement along the Sierra Madre Occidentale into the Mogollon Highlands, resulting in the presence of early maize at sites such as Bat Cave, Tularosa Cave, and Jemez Cave in upland settings above 6,000 feet. It then spread onto the Colorado Plateau via contact with other foragers. This early maize was thought to be adapted to high elevations and early frosts, and these upland settings were thought to be important for early agriculture because they provided higher levels of soil moisture than was present in lowland settings, making them more conducive for growing early strains of maize. Later, after some genetic changes led to higher levels of drought tolerance, maize was introduced to lowland settings.

The highland corridor has been supported by a study of nuclear DNA from maize samples from McEuen Cave in central Arizona (see chapter 5) and from Bat and Tularosa Caves in the Mogollon Highlands (Fonseca et al. 2015). The DNA data suggest that the initial diffusion of maize likely occurred along a highland route with later gene flow from a Pacific coastal route around AD 1.

Wills (1988a) and Huckell (1990, 1995) were the first to propose that lowland riverine corridors also served as routes of transmission, arguing that maize spread along the floodplain and adjacent alluvial terraces of rivers that drained the north-south basins that connected northern Mexico and southern Arizona. Huckell linked the occupation of these waterways to periods of aggradation postdating the Middle Holocene. Thus multiple routes of transmission, both upland and lowland, may have been used, but the lowland route has yet to be confirmed by genetic data. Current maize dates show that the lowlands had maize as early as the upland settings, so irrespective of the initial entry point, these data indicate that maize spread quickly to all regions (Smiley 1994).

The idea of a lowland corridor was derived in part due to growing evidence for the occupation of lowland alluvial settings in the southern Southwest. During the 1990s, Cultural Resource Management projects along the Santa Cruz River and research at several buried sites along the San Pedro River in southern Arizona revealed deeply buried sites with very different characteristics from those observed in the upland maize-bearing sites thus far investigated (Huckell 1990, 1995; Freeman 1997a; Mabry 1998a; Gregory 2001). These sites had pithouses, large bell-shaped storage pits, hundreds of other pits, and a diverse material culture including shell, ceramic figurines, and a wide variety of chipped and ground stone artifacts (figure 2.2). Most of these sites were located along key reaches of

Figure 2.2. Late Archaic period features at the Santa Cruz Bend site. Photo by the author

the Santa Cruz River in the Tucson Basin and spanned the time period from 1200 BC to the introduction of ceramic vessels at around AD 200. In addition to the substantially different material culture present at these sites, backhoe trenching revealed evidence of irrigation, first identified at the Costello-King site in the Tucson Basin (Ezzo and Deaver 1998) and eventually shown to be present at many sites along the Santa Cruz River floodplain (Mabry 2002; Nials 2015). These sites were viewed as so dramatically different from Middle Archaic foraging adaptations documented in the Tucson Basin that Huckell (1995) called for renaming the Late Archaic period the "Early Agricultural Period" in the southern Southwest to further emphasize these differences.

The interpretation of these finds must be understood in the context of the times in which they were found. As noted, prior to this, many of the sites with early maize were in upland settings and primarily in rock shelters (Smiley 1994). The discovery of these large riverine pithouse "villages" was touted as a dramatic change in the way that agricultural beginnings occurred; in fact, Mabry (1997:1) referred to their discovery as "rewriting Southwestern prehistory." Instead of being a minor addition with minimal impact, cultivation was viewed as having a tremendous impact on local populations in southern Arizona, to the extent that cultigens must have been introduced not by interaction with other foragers but by migrating farmers (Huckell 1995).

Continued work at sites in both the northern and southern Southwest into the early 2000s further muddled our understanding of agricultural origins. Excavations at the Clearwater (Thiel and Mabry 2006a) and Las Capas (Whittlesey and Foster 2010) sites along the Santa Cruz River pushed the dates of the earliest maize back to 2000 BC, and these remains were associated with more ephemeral occupations than the later "Early Agricultural" remains. Early maize was also identified at the Old Corn site in the Zuni region of northwestern New Mexico dating to 2100 BC (Huber 2005)—this remains the earliest corn in the Southwest and was found associated with a seasonal foraging site containing chipped stone, hearths, and storage pits. These data indicated that maize was present in some portions of the Southwest by as early as 2000 BC and that the pithouses, large storage pits, and other characteristics thought to represent such dramatic differences from the Middle Archaic occurred some five hundred to one thousand years after this initial introduction. This has led to somewhat of a stalemate in our understanding of agricultural origins as we attempt to integrate these important finds into our existing understanding of how, when, and why agriculture came into the Southwest.

Current Models

Two models are currently used to account for the presence of agriculture in the Southwest: 1) the migration of farmers from northern Mexico or 2) the adoption of cultigens by indigenous hunter-gatherers via interaction with maize-using groups. The major tenets of each model are outlined below.

Migration Model

The migration model, based primarily on data from sites located in riverine floodplain settings that crosscut the Sonoran Desert and Basketmaker sites in the western portion of the northern Colorado Plateau, and on linguistic data, posits that migrant farmers from northern Mexico entered southern Arizona in pursuit of agricultural land and occupied the broad alluvial expanses adjacent to the major rivers in this region (Carpenter et al. 2002; Hill 2001, 2002; Huckell 1995; LeBlanc 2008). They then moved north, eventually settling on the Colorado Plateau (Berry and Berry 1986; Matson 2002; LeBlanc 2008), and then further interactions with other foraging groups, especially to the east, led to the spread of maize agriculture from these migrant settings. It is important to recognize that the migration model was developed during a time when the majority of early maize dates were in the 1000 BC range.

The migration model is associated primarily with early work by the Berrys (Berry and Berry 1986) and later Matson (1999, 2002) and by Jane Hill's linguistic work (Hill 2001, 2002). Berry and Berry, Matson, and Hill all view the introduction of agriculture to the Southwest as "big" movement—with farmers who were sedentary and fully dependent on agriculture moving into alluvial settings in the Southwest and either outcompeting or otherwise displacing hunter-gatherer groups.

Using the existing radiocarbon sequence, Berry and Berry (1986) argued that the occupation of the Colorado Plateau was punctuated and episodic, with periods of abandonment due to droughts followed by reoccupation by migrants. These episodes included depopulation during the Middle Holocene (Altithermal) followed by a Late Archaic occupation characterized by Gypsum contracting-stem projectile points and split-twig figurines from 3000–1000 BC. This occupation ended when groups moved out during a subsequent drought, followed by movement into the area by San Pedro Cochise immigrants from southern Arizona around 500 BC, with these groups developing into the Basketmaker II, an idea later expanded on by Matson (1999, 2002). Berry and Berry (1986:319) spell out their interpretation: "the introduction of maize farming was accomplished through sociocultural intrusion rather than through diffusion of agriculture to indigenous hunter-gatherer populations." This theory had a tremendous impact on researchers in the region, especially in the northern Southwest, despite the fact that the Berrys never discussed why this so-called intrusion occurred, nor did they identify any region in Mexico from which these early farmers came.

Matson's (1999, 2002) study of the introduction of agriculture to the northern Southwest led him to be one of the major proponents of the migration model. He initially argued that both models of migration and indigenous adoption could be valid depending on what portion of the Southwest a researcher studied (Matson 1991), but later, using data from Cedar Mesa in southeastern Utah, argued that Basketmaker II groups in the northern Southwest represented San Pedro Cochise migrants from the southern Southwest who were highly dependent on agriculture. He based his theory on the lack of Archaic period occupations in many portions of the northern Colorado Plateau and on distinctions in housing, projectile points, and textiles between Western and Eastern Basketmaker groups, arguing that Western Basketmaker groups were more similar to San Pedro Cochise than Eastern Basketmaker groups and thus represented migrating farmers, while Eastern Basketmaker groups represented the adoption of agriculture by indigenous hunter-gatherer groups from western groups. Isotopic data from burials indicated that Basketmaker II groups on Cedar Mesa were as dependent on agriculture as later puebloan groups (Coltrain et al. 2007), which supported the idea that heavy maize dependence was associated with the initial migration of farmers into the northern Southwest.

Although Matson asserts that connections between Western Basketmaker and San Pedro Cochise groups exist, they are only presented as figures and summary discussions (Matson 1999:12; 2006); nowhere are primary data on the relationships presented in detail. As will be discussed in chapter 5, San Pedro points are widespread across large portions of the Southwest and thus are not particularly useful identity markers (Roth 2014). Additionally, although Matson indicates that San Pedro points are similar to side-notched Western Basketmaker points, Geib (1996:62) has argued that Western (White Dog) Basketmaker points are more similar to Elko points used by Archaic period groups in the northern Southwest. Textile data from the earliest farming sites in the southern deserts, those in the Tucson Basin, are lacking, and Matson's textile similarities instead derive from cave sites in the Mogollon Highlands, which are considered by Wills (1988a; 1995) and others (e.g., McBrinn 2005) to represent the adoption of agriculture by local hunter-gatherers. House similarities are again broad scale; for example, houses at the Keystone Dam site in the Jornada region of southern New Mexico (O'Laughlin 1980) are similar to those in the Tucson Basin and to early Basketmaker structures on the northern Colorado Plateau, so this similarity may be tied to function rather than serve as a cultural marker. The reasons why farmers from the southern Southwest would move north over five hundred kilometers to inhabit a new region that by all accounts is marginal and risky for agriculture have not been discussed or explained. It is clear that we need more rigorous data to link San Pedro Cochise groups from the south to Basketmaker groups in the north if this migration model is to be considered tenable.

Hill (2001, 2002) has also been a strong proponent of the migration model. She argues that the initial migration of farmers into the Southwest was associated with the spread of Proto-Uto-Aztecan speakers from Mexico between 2500 and 1500 BC. She did not base this on the archaeological record, but on linguistic data associated with the terms used today by Uto-Aztecan speakers for maize and agricultural activities. Hill argued that these migrating groups first moved into southern Arizona, represented by early floodplain agricultural sites, and some groups later moved north and became the Western Basketmaker. She cites population pressure from growing populations in central Mexico as the reason for this migration.

Carpenter et al. (2002) have further supported this idea, arguing that Uto-Aztecan speakers migrated into southern Arizona between 3000–1000 BC following the amelioration of the climate after the Altithermal. They argue that the lower desert regions of the Southwest were completely abandoned during the Altithermal, leaving open alluvial zones for migrating farmers once wetter conditions prevailed. This proposed abandonment is based on the idea that Middle Archaic projectile points and associated occupations recovered throughout the southern deserts postdate the Altithermal, although recent data do not support this (see chapter 5). Both Hill (2001, 2002) and Carpenter et al. (2002) used the presence of sites in the Tucson Basin dating to 1200 BC with pithouses, storage pits, canals, and abundant maize to illustrate the initial migration of Uto-Aztecan speakers.

Hill (2002) and LeBlanc (2008; LeBlanc et al. 2007) have cited genetic evidence to support the idea of a Mesoamerican origin for early farmers in the Southwest, but this is also not supported by current data (Merrill et al. 2009). Mitochondrial DNA studies show marked differences in female lineages between the Southwest and Mesoamerica (Malhi et al. 2003; Kemp et al. 2010). Although Hill (2002), LeBlanc (2008), and more recently Sliva (2015) and Heidke (2015) suggest that this may indicate that only males migrated and brought maize agriculture to the Southwest, this seems highly unlikely given the abundant ethnographic and archaeological data highlighting the significant role of women's knowledge of plants, processing, and storage activities in successful agriculture (Roth 2006). This interpretation of male-only migration also raises the question of why only men would migrate if population pressure was the main causal factor for migration. Byrd (2014) has documented cranial differences in skeletons from early agricultural sites in the Tucson Basin that she sees as representing male exogamy. Intermarriage may therefore account for the presence of non-local males, and this would be consistent with the maintenance of social networks that were so crucial to hunter-gatherers in arid environments (see chapter 3).

This debate ties into an even broader theoretical issue, which is the role of migration in past cultural change. Early Cultural Historical archaeologists emphasized migration and diffusion as the major forces in bringing about culture change. Later processual archaeologists, in part seeking to disassociate themselves from the earlier cultural historical approach, largely ignored migration as a possible mechanism of change, focusing instead on environmental and technological variables. More recently, archaeologists are returning to migration as a potentially powerful force in culture change, examining both the causes and consequences of the decision to migrate (Anthony 1990; Cabana and Clark 2011).

Rouse (1958) was the first to formally address the archaeological correlates of migrants, arguing that the process of migration would result in distinct patterns that would be apparent in the archaeological record. These include, but are not limited to, 1) a "donor" region from which the migrants moved; 2) a compelling reason for movement, especially if migration involved the movement of more than individual households; 3) a place to move to—either an "empty niche" (Haury 1976), an area with available resources with which the migrating population had ties, or evidence of the displacement of local groups manifested in evidence of conflict and competition; 4) genetic and/or linguistic data that indicate the influx of a new population; 5) differences in architecture from local patterns; and 6) differences in material culture from local patterns. These archaeological signatures may vary depending on the type and extent of migration, but the idea of migration presupposes the idea that migrating groups represent separate identities.

Recent work on migrations in the American Southwest, while focusing on later prehistory, can provide insights into the migration process. Migration often occurs on the household level (Bernardini 2005; Cameron 1995; Duff 1998) with households moving in with groups with whom they have had previous connections either via kinship, trade, or other alliance mechanisms. Often this small-scale migration results in the assimilation of the migrant groups, with some archaeological evidence of the introduction of new technological ideas or styles (Stone 2003). Even large-scale migrations of families often occur along previously established trade or kinship lines. In these cases, migrant groups may remain archaeologically distinct in terms of architecture, ceramic styles, and ceremonial practices and live alongside local populations (cf. Clark 2001; Woodson 1999).

The broad-scale migration of farmers proposed by proponents of the migration model should be readily apparent in major differences in occupation between local hunter-gatherers and immigrant farmers, with farmers bringing different architecture and technology, especially as it relates to agricultural production. The current archaeological data indicate that this is not the case, and thus it has become apparent that the archaeological data are more complex than the migration model allows. Sites such as Clearwater in the Tucson Basin (Mabry 2006), Old Corn in the Zuni region (Huber 2005), and Simmons's (1986) data from the San Juan Basin show that early maize was introduced in a variety of settings at about the same time (ca. 2000 BC), but it arrived to what appear to be very typical Archaic period hunter-gatherers rather than distinct farmers. Movement of farmers into other areas occurred later, particularly on the northern

Colorado Plateau, but this represents a different pattern of "migration" from the initial movement of farmers into the Southwest and needs to be evaluated separately.

Indigenous Adoption Model

The second model proposed to account for the introduction of agriculture into the Southwest is that indigenous hunter-gatherers adopted agriculture. Researchers who propose this model have argued that the adoption occurred for varying reasons depending on the environmental setting, population density, and the nature of the Archaic period occupation (Hogan 1994; Irwin-Williams 1973; Minnis 1985, 1992; Parry et. al 1994; Roth and Freeman 2008; Whittlesey and Ciolek-Torrello 1996; Vierra 2008; Vierra and Ford 2006; Wills 1988a,b, 1992, 1995).

Irwin-Williams's (1973) pioneering work on the Archaic period centered on the idea that local Archaic period groups adopted maize, followed by a long period of casual maize use prior to heavy reliance on agriculture. She stressed that maize use was tied to regional population densities and environmental factors, and this idea has conditioned the way that many researchers who see an *in situ* adoption view the transition to agriculture, especially around the areas of the Southwest where Irwin-Williams worked such as the Rio Puerco and northern Rio Grande regions (Post 2002, 2013; Vierra 2008; Vierra and Ford 2006). Vierra (2008) argues that demographic patterns observed in Archaic populations who adopted maize reflect long-term repositioning on the landscape in response to subsistence change, which could in some circumstances mimic the migration of farming populations.

Wills has published extensively on this topic (1988a,b, 1992, 1995), focusing particularly on the adoption of maize by Archaic period foragers in the Mogollon Highlands. He cites evidence that foragers and early farmers exhibit technological and subsistence continuity, which he sees as indicating that migration was not the means for the introduction of agriculture into the Southwest. He has also argued (2006:124) that "the underlying economics of foragers and farmers were fundamentally similar . . . allowing tactical adjustments rather than strategic change." He notes, however, that the role of early cultigens varied depending on local ecological conditions and sees a divergence between the northern and southern Southwest by 800 BC (Wills 1995:219).

Merrill et al. (2009) have argued that maize was initially introduced to the Southwest after foragers in contact with groups from northern Mexico passed it from group to group using previously existing social networks. They note that no artifacts or features of any kind point to in-migrating Mesoamerican farmers, instead citing the abundant evidence that exists for local continuity in the archaeological record (Merrill et al. 2009:21024). Like Wills (1992) and Minnis (1992), they propose that foragers incorporated maize into their diets as a supplement to their broad-spectrum subsistence base. They argue that the linguistic data used by Hill (2001, 2002) to document a migration of Uto-Aztecan speakers can be better interpreted as resulting from a very early (ca. 7000 BC) spread of Proto-Uto-Aztecan speakers from the Great Basin, not from Mexico. The later separation of northern and southern Uto-Aztecan speakers after maize was introduced accounts for the lack of shared terms for maize by these groups.

New Approaches to Agricultural Beginnings

New dates from sites with early maize in the Southwest have challenged the migration model because the sites do not exhibit the substantive changes observed at later post–1200 BC sites that have been used to argue for the migration of farmers into the region. They call for decoupling the initial use of maize, which is early, widespread, and associated with seasonal foraging adaptations and a degree of residential mobility, from the maize-bearing occupations used by most researchers to argue for a migration, which date anywhere from five hundred to one thousand years after the initial arrival of maize. Thus the Basketmaker II movement of farmers onto portions of the Colorado Plateau at 500–300 BC cannot be viewed as resulting from the same mechanisms and processes as the initial movement of maize onto the Colorado Plateau, which, if dates from Three Fir Shelter are corroborated, occurs 1,500 years earlier. We need to develop more sophisticated models for examining both the initial entry of maize and the subsequent movement of farmers if we are going to understand the cultural processes associated with agricultural beginnings in the Southwest.

Why Cultigens?

If we disassociate the initial use of maize from these later occurrences, then understanding agricultural beginnings requires addressing the reasons for the adoption and use of this new dietary component by hunter-gatherers. The challenges of interpreting the reasons for the adoption and use of maize are similar to those faced by researchers attempting to develop models to explain the initial process of plant domestication worldwide—we are not dealing with a unified phenomenon—and there appear to have been multiple interrelated factors involved, and these varied based on population density and environmental setting (e.g., Gremillion and Piperno 2009; Rowley-Conwy and Layton 2011).

The reasons proposed to account for the adoption of maize by Southwestern hunter-gatherers generally revolve around its predictability, storability, and the ability to manipulate (and thus expand) productivity. These models usually consider maize as a supplement to the diet but focus on its impact on hunter-gatherer adaptations and the subsistence reorganization that maize cultivation brought about. Most researchers discuss reasons that are region specific and recognize that the reasons for the adoption of maize in the northern Southwest may have been different from those in the southern Southwest.

Predictability and Productivity

Wills's (1988a) landmark study of Bat Cave was the first to argue that the adoption of maize by Southwestern foragers was a conscious decision. He emphasized the significance of this decision-making process, arguing that the incorporation of domesticates would be used to "sustain the on-going system instead of changing it" (Wills 1988a:36), a pattern consistent with the fact that hunter-gatherers are generally risk averse (see chapter 3). Wills focused on the role of cultigens in reducing overall subsistence uncertainty and increasing the density of food resources. Diversification of the subsistence base is a common response to uncertainty by prehistoric foragers worldwide (Halstead and O'Shea 1989) and the adoption of maize can be understood as part of this response.

Maize cultivation would have increased the stability and predictability of the resource base for a number of reasons. It would provide carbohydrates and calories that could be used to combat subsistence stress during periods such as the early spring when stored supplies were depleted, local wild resources were scarce, and local game was lean (Speth and Spielmann 1983). It would also increase resource density, which could not be done by manipulating wild plants. This increase in density could offset food shortages during years when wild resources were not abundant such as when piñon masts were low, and it could also provide more food for growing populations in regions where regional population densities were increasing. The ability to manipulate productivity levels was a major difference between cultigens and wild resources, and L. Huckell (2006) shows that it was one of the first changes that we see in maize after it was initially used, as early farmers apparently selected for increased cob row during the first one thousand years in which maize was grown. However, Diehl (2005a) has noted that although the ability to increase productivity may have been a motivation for initially adopting cultigens, early farmer-foragers did not manipulate maize to the extent that would be expected if this was a primary goal of cultivation (see also Wills 1992). Winterhalder and Goland (1997:133) argue that even a marginal change in foraging efficiency can have significant economic ramifications, and small increases in resource productivity would enhance the buffering capacity of any resource; this may be the case with early maize use in the Southwest.

Wills (1988a:193) emphasized that the predictability of maize made it particularly attractive to foragers because of the seasonal food security that it provided. Vierra (2008) has stressed that maize played an important role as a back-up strategy during droughts or seasonal shortfalls in the Rio Grande region, and Diehl (1997:263) has argued for a similar role for maize in the southern Southwest, with maize adopted as a risk-minimization strategy to counteract seasonal shortfalls by providing a stable storable surplus of high-calorie food.

Storage

Storability has been proposed as another major reason for the adoption of agriculture, as stored maize would provide important nutrients during lean periods such as winter and would serve as a buffer to counteract fluctuations in wild resources. Storage technology is the major technological change associated with the earliest maize cultivation in the Southwest, as bell-shaped storage pits are present at most of the sites with early maize (figure 2.3), and storage technology was probably adopted along with the maize itself.

The impact and significance of the use of stored maize likely varied between the northern and southern Southwest. Dello-Russo (2006), following earlier researchers such as Hunter-Anderson (1986), has argued that the ability to store maize was one of the key aspects leading to its integration into the diet in the northern Southwest, where stored maize would have enabled overwintering in response to predictable seasonal shortages. In the southern Southwest, the ability to store maize through lean periods and counteract seasonal shortfalls would also have been important, but storage may also have been important because population densities were higher along the major waterways of the Sonoran Desert. The presence of large-capacity storage pits at many riverine sites occupied after the initial adoption of maize illustrates the significance of storage technology to these

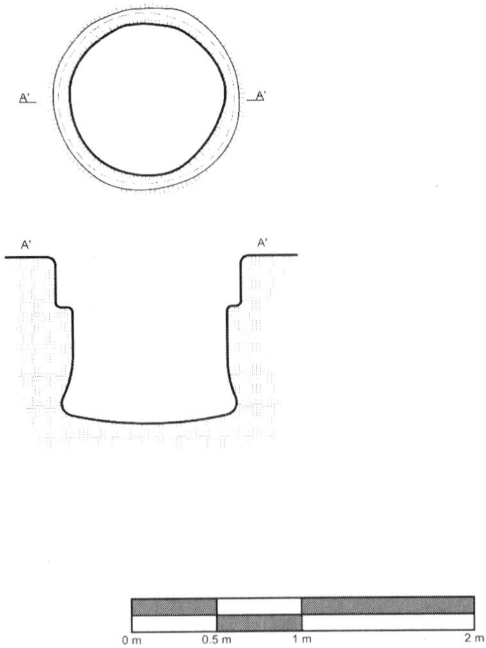

Figure 2.3. Example of bell-shaped storage pit. Figure by Ben Van Alstyne

groups (Huckell et al. 2002). Storage pits could also have been used for storing wild resources to provide additional buffering of seasonal resource stress; this is supported by the recovery of eight hundred carbonized amaranth seeds in a large storage pit at the Late Archaic period Milagro site in the Tucson Basin (Huckell et al. 2002).

Despite the importance of storage to these early farmers, Diehl and Waters (2006) argue that pit storage prevented the intensification of maize use and further manipulation of maize, citing evidence of storage losses up to 30 percent that made it too risky to invest heavily in maize. Instead, groups used it as a component of a broad-spectrum diet until the advent of ceramic storage technology.

Compatibility with the Existing Subsistence Regime

The initial adoption of agriculture took place in areas where it was compatible with the existing subsistence and settlement system and did not require substantial changes in resource procurement scheduling or place additional labor requirements on hunter-gatherers (Wills 1992, 1995). Maize cultivation still required labor investment in soil preparation, planting, tending, harvesting, and processing, so hunter-gatherers had to make decisions about the costs versus benefits of devoting labor to these tasks. It is thus not surprising that we see the earliest maize in settings where it would have fit in with seasonal resource procurement and could be easily integrated (Minnis 1992; Wills 1992, 1995; Roth and Freeman 2008). If groups were also manipulating wild plant resources, as discussed in chapter 1, then the addition of another species with similar requirements could have been accomplished with very little overall impact on the subsistence regime. In addition to the direct benefits of maize, cultivation would also have secondary benefits as the disturbed soil from tilling would provide excellent habitat for weeds such as chenopodium, and gardens would have attracted game.

Processing costs of maize also need to be considered when evaluating its contribution to the diet and the reasons for its adoption. Mabry (2005b:141) notes that processing costs including shelling, pounding, and grinding could be two to three times the energy costs of field labor. Adams (2005, 2015) has further argued that these processing costs led to the initial use of immature maize that was processed by roasting, boiling, popping, or parching rather than by grinding. Maize stalks may have been cut and eaten or pounded for juice to take advantage of their high sugar content (Smalley and Blake 2003). Adams (2015) notes, however, that the advantage of using immature maize, and thereby reducing processing costs, is offset by the fact that it cannot be stored as long as mature maize.

Despite the processing costs and generally low overall productivity of early maize relative to later varieties, Barlow (2002) has shown that maize would rank higher in terms of energy returns than many wild plant resources, but still significantly lower than large game. This may account for its initial adoption in areas with lower overall risk due to the presence of reliable and stable wild resources, which could have served as back-up resources when crops failed (Diehl 1997).

In different portions of the Southwest, population density may have been another factor influencing the adoption and subsequent use of maize agriculture. Pryor (1986), in looking at the origins of agriculture worldwide, has argued that the presence of agriculture is directly tied to population density, and Redding (1988) has argued that both local population growth and inter-annual fluctuations in the resource base led to domestication to supplement the diet and counteract the unpredictability of the resource base. Demographic reconstruction is always difficult using archaeological remains, and Kohler et al. (2008) have established that there was not a dramatic demographic shift in the Southwest associated with agriculture until AD 500, after the introduction of beans, squash, and ceramic technology, but it appears that population densities were higher in alluvial settings of the Tucson Basin and perhaps the San Pedro River Valley than elsewhere in the Southwest shortly after the initial appearance of maize. There are no indications that population pressure was a causal factor in the initial adoption of maize, but it may have played a role in the later intensification of agriculture and in the movement of farmers into new areas.

"Sweetness Hypothesis"

Finally, some researchers have suggested that because of the low productivity of early cultigens, other social and demographic factors must be considered to account for their use and spread (Bowles 2011). Smalley and Blake (2003) have argued that maize was initially domesticated in Mesoamerica because of its sweet stalks rather than the kernels and cobs. The "sweetness hypothesis" proposes that the stalks were the main target for consumption because of both the sugar contained within them and the sweet juice produced by them, which could be fermented and used during social gatherings. Thus, social rather than economic factors were most influential in the early domestication of maize.

While this theory is intriguing to account for the initial domestication of teosinte in Mexico, and certainly Archaic period foragers in the Southwest would have been amenable to adopting a source of sugar, the cultivation of maize for sugary stalks does not explain the storage technology that co-occurs with the introduction of maize in all areas where it is initially adopted in the Southwest. The consumption of maize stalks and the use of juice from the stalks would not necessitate this level of storage. If fermentation was a key factor influencing maize use, then some kind of vessel for fermentation would be necessary; gourds could not have been used because the acids would have eaten through the container during the fermentation process. We would thus expect stone- or ceramic-vessel technology to have spread quickly into regions where maize was first used, and we do not see this happening. Although it is highly likely that maize stalks were consumed, it appears that the kernels were the major target for consumption by early farmers in the Southwest, and it is likely that the carbohydrates and calories in them were a fundamental reason for their consumption and use.

Agricultural Strategies

Mabry (2005b) and Mabry and Doolittle (2008) have detailed the array of agricultural techniques used by early farmers in the Southwest and have documented that these were tied to specific environmental circumstances and reflect both diversity and flexibility in agricultural practices. They note that these techniques vary in terms of productivity, risks, and labor requirements, all of which would have been weighed by groups before both adopting maize and intensifying production.

The techniques used by the earliest farmers included 1) floodwater farming, with fields located in areas that flooded seasonally; 2) water-table farming, which involved placing fields in areas with high water tables which Mabry (2005b:125) refers to as "sub-irrigation"; 3) runoff or *ak chin* farming, which involved the use of natural gradients and farming along seasonal streams; and 4) irrigation. These strategies could have been practiced in alluvial settings throughout the Southwest, and they have been documented in the southern Southwest and on the Colorado Plateau between 2100 and 1500 BC (Mabry 2005b).

Mabry (2005b) found different trade-offs in risk and labor associated with each of these techniques and ranked them based on these differences, arguing that water-table farming would occur first because of the low risks, low labor requirements, and high yields, followed by floodwater farming, and then runoff and irrigation farming. Floodwater, water-table, and runoff farming required permanent, fixed plots but were sustainable systems while irrigation required more open land and shifting fields (Mabry and Doolittle 2008:112). Mabry (2005b) notes that the choice of techniques would be conditioned by elevation, with irrigation more likely to be successful at lower elevations. Groups may have practiced multiple strategies to ensure crop yields and lower risk. Dry farming, which involved locating fields in moisture-retaining soils such as dunes and terraces, occurred later, most likely after all of the other agricultural niches were filled (Mabry and Doolittle 2008), which may account for some of the later movement of farmers on the Colorado Plateau (see chapter 4).

Evidence of irrigation has been found on the floodplain in the Tucson Basin by 1500 BC (Mabry 2002) and in the Zuni region of northwest New Mexico by 1200 BC (Damp et al. 2002), and Mabry (2005b) argues that it was independently developed in both Mesoamerica and the southern Southwest. Irrigation farming would have required a specific amount of labor that likely resulted in labor and scheduling reorganization to make it successful (Mabry 2002). Irrigation is also cross-culturally associated with sedentism, implying a certain degree of residential stability by groups practicing it (Mabry 1996).

Intensification

Intensification is a difficult concept to define and quantify (see overview by Morgan 2015) and in the Southwest it has usually been intertwined with the concept of agricultural dependence. Binford (2001:188) defined intensification as "tactical or strategic practices that increase food per unit area." This definition encompasses an array of tactics that lead to increased productivity, including changes in labor input to different subsistence pursuits (including cultivation), technological innovations for more efficient procurement and processing, and storage. Intensification was often accompanied by changes in settlement strategies, usually increased sedentism in response to scheduling demands, labor organization, and technological options such as storage. The factors influencing the decision to intensify resource procurement or invest in food production vary, ranging from demographic and climatic factors to the presence of abundant resources allowing risk-taking strategies.

Subsistence intensification worldwide has been characterized using archaeological and artifact data, including evidence of storage facilities, structures, population increase, and new processing and agricultural technologies. Intensification in the Southwest is usually tied to agricultural production, and evidence for intensification varies regionally but generally includes these same kinds of archaeological remains, including larger sites with more substantial structures, storage, and sometimes canals or terraces. Changes in ground stone have been argued to be associated with agricultural intensification with increased mano size viewed as correlated with increased grinding efficiency (Hard 1990; Hard et al. 1996; Mauldin 1993), but Adams (1999) has argued that these changes are related to food-processing strategies rather than the method of food procurement, so they do not necessarily correlate with agricultural production. The variability in agricultural intensification observable over time and space across the Southwest is likely related to the varying circumstances associated with the initial use of maize agriculture. It is addressed in subsequent chapters for specific regions, but region-wide differences in intensification should be a topic for future research as we begin to understand the role of agricultural beginnings in later cultural change.

The Impact of Maize Agriculture

The existing archaeological evidence indicates that maize was initially introduced to the Southwest during a period when Middle Archaic period foragers were practicing relatively broad-spectrum subsistence strategies tied to seasonally available resources, but had already transitioned to a focus on plant resources. This led to different degrees of sedentism in different settings. Some groups remained highly mobile, essentially "mapping onto" the seasonally available resources (Binford 1980). This is especially apparent in Middle Archaic groups in areas with dispersed water sources and patchy resource distributions such as the San Simon region of southeastern Arizona (Gilman 1997). Other groups shifted to what has been characterized as a logistical mobility strategy (Binford 1980), where they occupied certain areas with abundant resources for longer periods and exploited surrounding resource zones with forays to obtain specific resources, indicated by the presence of base camps with evidence of repeated occupation surrounding by specialized procurement sites. It is within this context that agriculture was first introduced.

Minnis (1985:310) has characterized the initial adoption of cultigens as a "monumental non-event with little immediate impact on native human populations," arguing that cultigens were incorporated into a foraging economy with dependence occurring substantially later, circa AD 200 in the northern Southwest. While this may be true of the *initial* use of maize, the impact of agriculture varied throughout the region.

As we look across the Southwest, it is clear that very different levels of investment in agriculture are present, almost from the start. In the Sonoran Desert, canal irrigation dating as early as 1500 BC led to a greater investment in cultivation, at least in terms of the labor invested in canal construction and maintenance and in the use of large storage pits. These factors suggest higher levels of sedentism during the Late Archaic period in this portion of the Southwest. The productivity of maize was not particularly high, even with the labor invested in irrigation, but substantial efforts were made to provide suitable land and water to these cultigens. Phillips (2009:697) has noted that labor costs would make irrigation practical where it was "as easy as it was productive." The impact that this had on population densities was considerable, as indicated by the large number of features found on the floodplain of the Santa Cruz River at later Late Archaic period sites (see chapter 5). The role of maize agriculture in this entire process appears complicated, and it is perhaps irrigation that led to the most substantial changes, but the differences are clear and important and have implications for the transition to Ceramic period occupations.

In contrast, in other portions of the Southwest such as the northern Rio Grande and Chihuahuan Desert regions, groups exhibit more casual use of maize (Miller 2016; Vierra and Ford 2006). This is manifested archaeologically in the addition of cultigens to a primarily seasonal foraging adaptation with the addition of storage pits enabling overwintering representing the most substantial change. The demographic consequences of this kind of supplemental use of maize are less dramatic than those observed in the southern Southwest, but still the addition of maize agriculture would have impacted how these forager-farmers used the landscape. There were several factors that were necessary for maize to succeed, and cultivation required some labor investment on the part of the foragers, which in some cases resulted in scheduling conflicts that had to be dealt with using either changes in the division of labor or in changes in the focus of subsistence pursuits (Wills and Huckell 1994). Wills (1988a) challenges the idea that maize could have been left untended for long periods, arguing instead, using several ethnographic examples, that groups had to regularly monitor fields to ensure any level of success. Still, the levels of investment vary across time and space in the Southwest. Thus we see regional variation not only in the initial use of maize but also in the long-term consequences of maize use tied to environmental and demographic variables.

Foraging Behavior and Agricultural Beginnings

3

As discussed in the previous chapters, the transition from hunting and gathering to farming in the Southwest initially involved hunter-gatherers adopting cultigens and later either intensifying production, sometimes resulting in the movement of farmers into new areas, or maintaining a stable long-term forager-farmer adaptation. In order to more fully understand and examine potential hunter-gatherer behavioral and technological responses to the introduction of cultigens, it is useful to examine ethnographic and ethnohistoric data on arid-land hunter-gatherers to assess the kinds of responses to be expected of these groups and the factors that played into their subsistence choices. The position taken here follows Diehl (2005b) and Minnis (1992), who argue that the earliest farmers in the Southwest did not immediately become heavily dependent on cultigens but instead incorporated them into existing foraging adaptations, with varying levels of intensification occurring based on the technology used in agricultural production, population densities, and the environmental setting. Diehl (2005c) refers to these groups as "farmagers" to illustrate this point.

Abundant ethnographic data are available on contemporary hunter-gatherers, but the challenges in using these data have been detailed by many cultural anthropologists and archaeologists (Ames 1999; Kelly 2013). By the time ethnographers began gathering data on hunter-gatherers, they had already been significantly impacted by contact with other groups, some to the point of completely changing their overall subsistence strategy (e.g., Pygmy groups in close contact with Bantu in Africa). Additionally, currently many foragers live in marginal locations such as the Australian

Outback and Kalahari Desert, where foraging adaptations are extreme compared to those faced by many prehistoric hunter-gatherers in more resource-rich settings (but see Porter and Marlowe 2007). There is therefore no one-to-one correlation between prehistoric and ethnographic hunter-gatherers because of the host of changes that have occurred during modern times. As with the use of any analogy, it is important to couch this use in the development of models of behavior and hypotheses that can be tested with the archaeological record rather than as a "window to the past."

Kelly's (2013) work on the "Foraging Spectrum" provides a strong foundation for using ethnographic data to model prehistoric foraging behavior (figure 3.1). He focuses on the long-term behavioral variability present in these hunter-gatherer groups. Both he and Binford (2001) have noted that variation in environmental settings and regional population densities significantly impacted foraging behaviors, leading to sometimes substantial variation in all aspects of hunter-gatherer society, including subsistence, mobility, and social complexity. Binford (2001) identified seven different "system states" among ethnographic hunter-gatherer groups that he tied to specific environmental and demographic variables, primarily ef-

Figure 3.1. Example of Hadza foraging activities. Photo courtesy of Alyssa Crittenden

fective temperature and population density. He argued that the conditions under which these system states are found in hunter-gatherer populations can provide insights into variation in prehistoric hunter-gatherer populations and may help explain change over time (Johnson 2014:6). In general, hunter-gatherer groups akin to Southwestern foragers at low latitudes exhibit more dependence on plant resources, a pattern consistent with the Southwestern Archaic period archaeological record.

This chapter examines data on ethnographic and ethnohistoric hunter-gatherer adaptations to arid environments, and specifically those in the Desert West, where reliable water sources and associated resource density and distribution structured hunter-gatherer subsistence and land use. Overall regional populations were not high, and social networks played an important role in hunter-gatherer behavior and interaction. Environmental and social factors influenced foraging activities in this setting, and foraging behavior cannot be understood without considering both.

Data on Great Basin hunter-gatherers can be used to model the issues and challenges faced by Southwestern Archaic period groups (Fowler 1992; Kelly 1964; Kelly and Fowler 1986; Liljeblad and Fowler 1986; Steward 1933). No ethnographic data on hunter-gatherers in the Southwest are available, as by the time ethnographers entered the Southwest, all of the native groups were using agriculture to some extent. However, a plethora of data are available on wild-plant use and hunting strategies used by various farming groups that can provide information pertinent to examining plant and animal use in the past (Bean and Saubel 1972; Bell and Castetter 1937; Castetter and Bell 1937; Castetter and Underhill 1935). The approach taken here is to examine how Archaic hunter-gatherers likely behaved prior to the introduction of maize and the factors that would have influenced their response to cultigens. To address this, general patterns of behavior observed in these ethnographic groups and gleaned from other studies of hunter-gatherers worldwide are used to provide a framework of foraging behaviors that I argue would have most significantly influenced prehistoric hunter-gatherer behavioral responses to the presence and subsequent introduction of maize and the processes that underlie the decision to adopt (or not to adopt) maize agriculture.

Hunter-Gatherer Behavior

Hunter-Gatherers Are Generally Risk Averse

A large body of data exists on responses to risk by ethnographic hunter-gatherers worldwide (Fitzhugh 2001; Cashdan 1990). In general, hunter-gatherers avoid risk by using a range of strategies to deal with any form of subsistence or social stress. Risk avoidance behavior is heavily predicated on the available options for responding to stress. As Halstead and O'Shea (1989) have outlined, forager responses to subsistence stress included mobility, increasing diet breadth, intensifying the use of particular resources and resource zones (often accompanied by storage), and sharing/exchange networks. Risk-taking behaviors tend to occur primarily under two opposing situations—either during periods of abundance or during periods of extreme subsistence stress.

Mobility is the most commonly used option, but it is predicated on the fact that there are places to move with available resources and that social networks allow for it. Other buffering strategies may take effect in situations where climatic or social situations do not allow for mobility (O'Shea 1989). These options include expansion of the diet to include resources not generally exploited that may require additional labor in harvesting and/or processing, storage to offset seasonal shortages, and either the development of or increase in the use of exchange networks. Storage is generally rare among ethnographic hunter-gatherers and it is not commonly found in the archaeological record associated with hunter-gatherers, although small-scale storage may have been done in hide bags, baskets, and cache pits. Pit storage technology in the Southwest appears to be uniquely tied to the introduction of cultigens.

Ethnographic hunter-gatherer exchange networks are based primarily on kin networks or on economic networks for access to resources that were not available in particular areas. Of particular interest for the introduction of cultigens to the Southwest, these social networks may have been established early and provided routes for the movement of both physical remains (e.g., cultigens) and ideas (e.g., cultivation, storage). Interaction with groups familiar with the process of cultivation likely facilitated the transmission of cultigens, and the knowledge passed with the cultigens was essential to their successful integration into new areas. Food exchange networks are only operational within particular distances to make them feasible, but exchange of goods and kin networks would also have served to create relationships that could ensure either temporary residence or other forms of aid during lean periods.

In general, hunter-gatherer societies were and are quite stable and resist change when possible. However, because of the nature of hunter–gatherer lifeways, these groups are also generally quite flexible and adaptable in the face of change. Stutz (2012:28) refers to this as "cultural resilience." He notes that it is a key factor in hunter-gatherer societies today with groups facing challenges of increasing globalization and participation in the world economy, but maintains that this concept can be meaningfully extended to the past. Cultural resilience would account for the adaptability that we see in Archaic-period hunter-gatherers and would have influenced their response to environmental and social change. The long-term stability in the diet documented by Diehl (2005b) and Wills (1995) in the southern Southwest after cultigens were introduced illustrates that long-term stability was present within punctuated periods of change.

Hunter-Gatherers Know Their Environment

Hunter-gatherers had knowledge of their environment that is sometimes difficult for many Western-trained researchers to understand. They were well versed in the seasonal patterns of movement of game and the availability of wild plants and planned for this seasonal variation. They knew when game would be at its fattest and when it would be stressed, and they knew the best hunting spots. They knew the best time to eat the shoots and ripe portions of plants; knew how plants would respond to heavy, light, or no rain; knew techniques, either by innovation or sharing, of how to prepare otherwise inedible plants like acorns into edible food; and knew what other resources could be used if their regular fare was not available. They also recognized that inter-annual variation in resource productivity would occur and had mechanisms for dealing with it. They might not always be able to predict the environment, but they could predict when the most difficult times of the year in terms of resource availability would be. Great Basin researchers have noted the impact of year-to-year variation in the productivity of piñon (Steward 1938; Sutton 1984; Thomas 1972), but ethnographic hunter-gatherers were aware of this and planned for it, usually through movement and social strategies such as food sharing and exchange (figure 3.2). Mithen (1990) discussed the importance of collective memory to prehistoric hunter-gatherers, which can be preserved in a variety of ways including storytelling and songs, and this was one way that environmental knowledge was maintained and passed on.

It is important to consider this when evaluating the adoption of cultigens, as these new plants were not introduced into a vacuum. When making the initial decision to adopt maize, the (likely) women, who were responsible for collecting and processing plant resources, would have obtained information from the groups they were adopting the plants from concerning the best places to plant, things to watch out for during the growing season, and the best ways to harvest, store, and process maize. This knowledge would have been shared along with the cultigens

Figure 3.2. Paiute Indians harvesting pine nuts (circa 1912). Paradise Valley Folklife Project collection, 1978–1982 (AFC 1991/021), American Folklife Center, Library of Congress

themselves, and would have been crucial for the successful introduction and integration of any new dietary item.

Ethnographic data also exist on strategies used by hunter-gatherers in the Great Basin and California to enhance resource predictability and density, with burning, transplanting, and seed broadcasting being the most common techniques noted in ethnographic accounts (Fowler 1996; Steward 1933; Shipek 1989). Burning has been found to have been especially important to groups in grassland settings (cf. Bird et al. 2008) and was a means to both enhance the productivity of plants and to clear an area of overgrowth to attract game. These strategies again were a crucial component of environmental knowledge, and like processing and seasonal information, would have been passed to others and through generations.

Calories Matter

Hunter-gatherers generally focus on easily procured, nutrient-rich resources when they are available—and develop strategies to increase caloric intake when they are not. Studies using concepts derived from Human Behavioral Ecology have enhanced our understanding of hunter-gatherer subsistence choice in living foraging groups and by extension, prehistoric hunter-gatherers (Bettinger 2009; Bird and O'Connell 2006; Elston and Zeanah 2002; Madsen and Schmitt 1998; Simms 1985). Diet breadth models developed from ethnographic data have been used by several researchers to examine Archaic period subsistence choice (Barlow 2002; Diehl 1997). The focus on nutrient-rich resources with minimal costs for procurement and processing usually leads to an emphasis on large game or fish in regions where they are prevalent and easily obtained, and on plant resources in regions where game and fish are less predictable and abundant than plant masts.

In the Desert West, foragers were faced with variability in the spacing and timing of resource availability, and this influenced diet composition, which varied seasonally. Ethnographic data from the Great Basin show that hunter-gatherers who lived in regions with low population densities, patchy environments, and seasonal variation in resource availability, such as the Southern Paiute, often practiced high mobility and maintained a focus on a broad-spectrum diet, including plant resources such as piñon and rice grass, along with game such as deer, antelope, and rabbits (Kelly 1964). Groups in other regions with more reliable and closely spaced resources, like the Owens Valley Paiute, often exhibited evidence of subsistence intensification (usually in the form of storage and/or plant manipulation using strategies such as broadcasting seeds, burning, and tending wild stands) and focused more on resources whose productivity could be enhanced with these techniques (Steward 1933).

If critical resources were readily available and populations were low, then hunter-gatherers' diets would generally remain stable. Subsistence reorganization in hunter-gatherers usually occurs when something changes; either the availability of particular high-ranking resources (e.g., drought) or when population density increased through population growth and/or packing due to reduced mobility options. This could result in expansion of the diet to include lower ranking resources, the development of new technology for more efficiently processing other resources, or the adoption of new subsistence items, like cultigens. As will be detailed in chapters 4 and 5, these factors likely influenced Archaic period diet choice.

Social Relations Are Key to Hunter-Gatherer Interactions

This concept is perhaps most crucial to our understanding of how and why cultigens were introduced to the Southwest when they were. Social relationships, and especially kin relationships, are essential to a hunter-gatherer band's viability and success. These kin and other social relationships form the foundation for most of the activities that take place day to day in hunter-gatherer societies, including child rearing, subsistence pursuits, tool production, and camp maintenance. Cooperation is a fundamental component of hunter-gatherer group interaction (Apicella et al. 2012). This cooperation is often based on kinship, but can also be based on friendship.

The role of cooperation and social interaction is most clearly seen in groups who aggregate seasonally to exploit highly productive but short-term producing resources such as piñon masts. These social aggregation sites served as more than just a means for exchanging products, but also provided a time for exchanging new ideas and many times provided opportunities for meeting mates from other groups, which further enhanced social ties. Stutz (2012:42) has argued that a coordinated social network would be especially effective in managing access to resource patches, which he argues was the case with dense stands of wild seeds in the Levantine Near East. This concept can also be applied to the Southwest, especially in regions such as the Mogollon Highlands and Colorado Plateau where seasonally abundant plants such as piñon were present.

Food sharing is a key component of social relations among hunter-gatherers and is important not only for the distribution of resources, but for maintaining social cohesion and solidifying social ties. Kelly (2013) notes that the ethos of sharing is enculturated in hunter-gatherer children at an early age. Sharing usually occurs along kin lines, especially of critical plant resources, but game is often shared among a larger social network. Many researchers have discussed the role of meat sharing by male hunters in achieving some level of status (Gurven and Hill 2009; Hawkes and Bliege Bird 2002; Marlowe 2007; McGuire and Hildebrandt 2005); the reasons behind this sharing are complicated and debate continues. For this study, the importance of sharing lies in facilitating and maintaining social relationships. Kelly (2013) has emphasized that food sharing can be seen as "social storage," a strategy for banking support for when needs arise.

Day-to-day decisions, such as where to gather or hunt, will be made by family groups, but because cooperation is such an important aspect of hunter-gatherer society, group decision making is also important. Major decisions, such as where and when to move, are done collectively and generally by consent (Kelly 2013). It is interesting to think about this in terms of the adoption of maize by Southwestern Archaic period groups. Elsewhere I have discussed that it is likely that women made the initial decision to adopt agriculture, given their association with wild-plant gathering and processing (Roth 2006), but the women also had to convince the rest of the group that this was a viable new subsistence strategy because it affected all of them. Some groups may have been more open to trying this new dietary item, perhaps because resources were seasonally abundant and they were able to incorporate something new, especially if it did not cause drastic changes in their scheduling or labor needs, or perhaps because kin ties with farmers who were already using maize provided them with information that indicated that it could help them through lean times. This may explain the sporadic yet widespread distribution of early maize, as some groups may have readily adopted it while others waited to see how it worked out for their kin, friends, and neighbors before they adopted it. This is also important in evaluating the intensification of maize agriculture, as the decision to construct canals or otherwise devote more labor to agricultural production would have been agreed upon by the entire group.

In a simulation study of hunter-gatherer cultural diversity and ethnicity, Del Castillo et al. (2014) note that the emergence of identity in hunter-gatherer populations was a consequence of mobility, territoriality, and interaction, as groups defined themselves as similar based on their degree of interaction. They also note that groups with similar identities are most likely to cooperate (Del Castillo et al. 2014:367). Using similar modeling, Barceló et al. (2014) argue that cultural boundaries likely occurred via detachment from previous homogeneous populations into groups with less and less contact. Thus, identity formation may account for both the variation and similarities observed in hunter-gatherer and early farming populations across the Southwest, especially after the initial introduction of maize. This idea will be explored further in chapter 6.

"Low-Level" Food Production

Rowley-Conwy and Layton (2011:854) have remarked that hunter-gatherer/farmers (or the farmagers discussed by Diehl 2005c), do not represent just an intensification of the hunter-gatherer niche, but a new kind of adaptation. Smith (2001b) has referred to this as "low-level food production," a concept that has merit for examining the new adaptation resulting from the adoption of maize in the Southwest. The basic premise of this idea is that that there is a "middle ground" between hunting, fishing, foraging, and agriculture with hunter-gatherer groups who exhibit some reliance on domesticates but are not "agriculturalists" (Smith 2001b:12). The plant manipulation observed in groups such as the Owens Valley Paiute and Kumeyaay in California fall into this category, and this appears to be a good description of the earliest farming strategies in the Southwest. Smith (2001b:24) and Greaves and Kramer

(2014:263) describe low-level food production as a potentially stable long-term economic solution; one of the interesting factors then becomes evaluating cases where it changed.

Applications to Agricultural Beginnings in the Southwest

Taken together, these data on foraging behaviors can be used to develop archaeological expectations for how hunter-gatherers would behave across the transition from foraging to farming in the Southwest. The general tendency of arid-land hunter-gatherers to be risk averse and the stability of subsistence strategies in the absence of environmental or population change support previous models of the introduction of agriculture indicating that the adoption and integration of maize was a conscious decision. As noted in chapter 2, the reasons for this subsistence change likely varied, but data gleaned from ethnographic hunter-gatherers indicate that it involved efforts to increase the predictability and productivity of the resource base. Maize was initially adopted in areas where it resulted in little change in the overall subsistence and settlement organization, again in keeping with what we know about hunter-gatherer behavior. Finally, it is important to recognize the role of social networks, sharing, and interaction in the adoption of maize. In the following chapters, data on agricultural beginnings in the northern and southern Southwest are presented, and the ideas discussed in this chapter are used to evaluate the reasons for the adoption of agriculture in each area. Differing responses and impacts of the adoption of maize are present throughout the Southwest, with an underlying adaptation akin to the low-level food production described above.

The Northern Southwest 4

IRWIN-WILLIAMS'S PIONEERING WORK in the Arroyo Cuervo region of northwestern New Mexico as part of the Anasazi Origins Project was the catalyst for our understanding of the Archaic period in the northern Southwest (Irwin-Williams 1967, 1968, 1973). She defined the Oshara Tradition as representing one of the major subdivisions of the Archaic in the Southwest. In turn, her work significantly influenced interpretations concerning agricultural origins in this region, as her framework for the development of the Oshara structured how many researchers working in the area see agricultural beginnings (Cordell and McBrinn 2012; Vierra and Ford 2006; Vierra 2008).

Irwin-Williams divided the Oshara into five phases and viewed these as representing an unbroken sequence from Archaic period hunter-gatherers through puebloan farmers, with changes such as the introduction of agriculture represented by gradual shifts in adaptation (figure 4.1). Her interpretation of this sequence set the stage for future interpretive models focused on an *in situ* agricultural transition.

Irwin-Williams (1973) saw the development of the Archaic from the preceding Paleoindian period as occurring at 6000–5500 BC, following a period of climatic drying that made big game hunting impossible. The Early Archaic phases, Jay and Bajada, represented a transition from specialized hunting to more generalized hunting and gathering. These groups were still highly mobile foragers. Three succeeding phases are pertinent to agricultural origins in this region—the San Jose (3200–1800 BC), Armijo (1800–800 BC), and En Medio (800 BC–AD 400) phases. Interestingly, these phases as defined by Irwin-Williams follow closely those defined

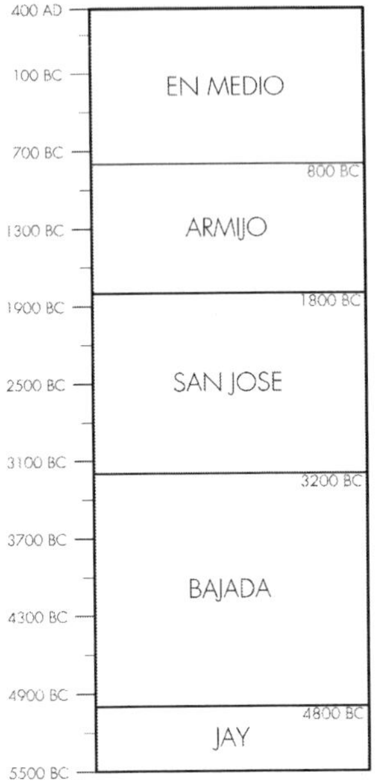

Figure 4.1. Oshara Sequence as defined by Cynthia Irwin-Williams (1968). Figure by Ben Van Alstyne

for the southern Southwest based on work in the Tucson Basin (see discussion of Middle Archaic, San Pedro, and Cienega phases in chapter 5). The San Jose phase, coincident with the Middle Archaic period, is characterized by increased site size and number over preceding Early Archaic period occupations. It witnessed the use of basin metates and small manos, documenting an increased focus on plant processing, although hunting remained a major subsistence pursuit. Ephemeral structures found at some San Jose sites suggest that settlement strategies were becoming more defined, and some sites in resource-rich areas were repeatedly occupied. Irwin-Williams (1973) saw the San Jose phase as representing the time period when hunter-gatherers in the Southwest began to exhibit more localized adaptations to specific microenvironments and emphasized that unique adaptations existed that were tied to local environmental conditions.

The succeeding Armijo phase was characterized by the addition of limited maize agriculture, with site locations shifting to floodplains on canyon floors and an increase in ground stone representing an emphasis on plant processing, including maize. In Irwin-Williams's framework, cultigens were added gradually to the diet and did not initially have a major impact on the groups who adopted them. She argued that some of the larger Armijo phase sites represented seasonal aggregation, but that in general groups continued to practice a degree of seasonal mobility. The En Medio phase represented the period when agriculture became more entrenched as a subsistence focus and population size and sedentism increased.

In the remainder of this chapter, the archaeological evidence for agricultural beginnings in the northern Southwest is presented, first in terms of evidence for Middle Archaic occupations that preceded the introduction of agriculture into the region, then the current evidence on the earliest agriculture, followed by a discussion of the period immediately following the introduction of agriculture into this area and the development of the Basketmaker II pattern. Cultural resource management projects have greatly expanded our understanding of the Archaic occupation of the "northern sector," especially in the Albuquerque and Rio Grande regions, and these data are incorporated where available. One note on semantics; the use of the term *Middle Archaic* in this chapter is consistent with the terminology used in the southern Southwest for the designation of hunter-gatherer occupations dating from circa 3500 BC–1500 BC and encompasses the San Jose and earliest Armijo phases as defined by Irwin-Williams (1973).

Environment

Simmons (2006) and Wills (1988b, 1989) have discussed the ecological differences between the northern and southern Southwest, arguing that environmental factors conditioned Archaic period adaptations in the two regions and consequently affected the adoption and integration of maize agriculture. Wills has emphasized that the northern Southwest is not particularly conducive to growing maize, given its aridity, lack of good-quality arable soils, and short growing season. The northern Southwest also exhibits much more inter-annual variation than the southern Southwest, which affected Archaic period subsistence choices and seasonal movements.

The role of the Middle Holocene drying period has been of major concern to many researchers working in the northern Southwest, but paleoclimatic data are not often robust enough from the entire region to determine the role of climatic change in events that occurred during the Middle Archaic period and its impact on hunter-gatherer adaptations. Research has documented a severe dry interval between 4500 and 3000 BC in the adjacent Great Basin that significantly impacted hunter-gatherers in the already dry regions of the California Deserts and central Great Basin, resulting in population movements to other regions with more reliable water and wild resource availability (Kennett et al. 2007; Louderbeck et al. 2011). It is not clear how this drying interval affected groups farther to the south and east, especially those on the Colorado Plateau. The distribution of Middle Archaic components in the northern Southwest, with some areas exhibiting substantial Middle Archaic occupations and others very limited occupations, may be a result of the differential distribution of reliable water and other resources and may represent the reorganization of foraging activities in response to climatic change. Both Wills (1988a) and Huckell (1996a) caution that the Middle Archaic drying period was not uniform across the Southwest, however, and in some areas such as the Mogollon Highlands, the expansion of grasslands may have actually increased resource potential. More fine-grained dating techniques other than the long time spans represented by projectile point styles and additional paleoclimatic data are needed to more fully understand the diversity of responses to environmental change represented in the Middle Archaic period record.

The Middle Archaic Period

One of the challenges faced by researchers in evaluating the transition to agriculture in the northern Southwest is the sporadic data on Archaic period occupations in this region. This is perhaps best exemplified by the fact that Kidder (1927) began the Pecos Classification (the sequence of cultural changes for the Anasazi [Ancestral Pueblo] developed using changes in architecture and pottery styles) with Basketmaker II, arguing that there was likely a Basketmaker I period that was not yet known due to the focus of fieldwork and research in the early 1900s on later Ceramic period occupations. Since then, many Archaic sites have been documented throughout the northern Southwest, but work at these sites has often been limited until recently. Much of the early work that was done at Middle Archaic period sites (or at sites with Middle Archaic period components) was done at rock shelters, and several researchers have discussed the problem associated with relying on sites that represent only part of a seasonal settlement system as proxies for interpreting past subsistence and settlement behaviors (Vierra 2008; Wills 1988a). Most of the recent research in the northern Southwest has been conducted as part of large-scale contract archaeology projects, and these data have substantially changed our views of both the Archaic period occupation of the region and the introduction of agriculture, in part because they have often focused on open-air sites. These data provide a framework for understanding Middle Archaic period foraging adaptations in this region and the role that they played in agricultural beginnings.

Middle Archaic occupations in the northern Southwest are characterized by a number of distinct projectile points style, which unfortunately are often poorly dated. San Jose points are side-notched points with expanding bases and are usually serrated (figure 4.2) (Irwin Williams 1973; Justice 2002). These points are sometimes classified as San Jose/Pinto, given their similarity to Pinto points from the Great Basin and other portions of the Southwest (Bayham et al. 1986). San Jose points are dated based on their recovery from radiocarbon-dated levels at sites in the Arroyo Cuervo region, thus placing them in the 3500–1800 BC range, and San Jose/Pinto points have been found at sites in southern Arizona dating to 3000–2500 BC (Bayham et al. 1986). However, Smiley (2002:32) reports on a San Jose/Pinto point from a site on Black Mesa associated with dates as early as 7000–6500 BC, suggesting its temporal range may extend back to the end of the Early Archaic period.

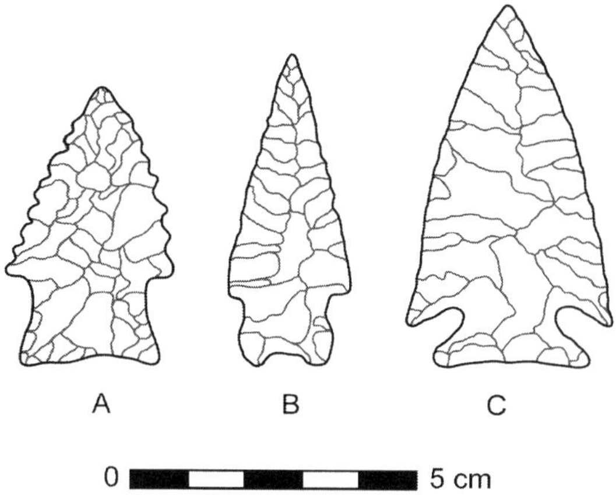

Figure 4.2. Middle Archaic period projectile points in the northern Southwest: A = San Jose, B = Pinto, C = Elko. Figure by Aaron Woods

Pinto points (figure 4.2) were first defined at the Pinto Basin site in the Mojave Desert and were described as thick points with narrow shoulders and incurved bases (Campbell and Campbell 1935; Amsden 1935). Numerous subtypes were later defined (Rogers 1939; Harrington 1957), leading Warren (1980:73) to conclude that "the status of Pinto typology at the present time might best be characterized as 'schizoid.'" Pinto points are also poorly dated; dates from the Mojave Desert place their use from 6500–2500 BC (Basgall and Hall 2000; Schroth 1994), but in general they have not been recovered from well-dated contexts in the northern or southern Southwest. Geib (2011:30) suggests that San Jose and Pinto points could have "considerably different temporal spans." Finally, Elko points, distinct corner-notched projectile points, appear in the northern portions of the Southwest during the latter portion of the Middle Archaic period (figure 4.2). This point type is widely distributed and encompasses a long time span, with dates ranging from 4200–300 BC in the Great Basin (Holmer 1986), and Elko points continued to be used into the Late Archaic period in the northern Southwest.

Middle Archaic Occupations

Early work on the Archaic in the northern Southwest identified a variety of different complexes, most of which are now subsumed under the more general Archaic (e.g., Desha Complex, Concho Complex). This section summarizes the evidence for Middle Archaic occupations in the portions of the northern Southwest where data are available (figure 4.3). As noted above, Irwin-Williams identified Middle Archaic components throughout the Arroyo Cuervo region based on the presence of San Jose points. Sites included base camps and specialized procurement camps with projectile points, a variety of other chipped stone tools, and often abundant ground stone.

Middle Archaic sites have also been found throughout the Rio Grande region (figure 4.3) dating back to early research in the area, but our knowledge of the occupation of this region has increased significantly as the result of work by Vierra and his colleagues (Vierra 2008, 2013; Post 2002, 2013; Vierra and Ford 2006, 2007; Vierra and Foxx 2010). Vierra (2008; Vierra and Ford 2006) argues that Archaic sites in this area, especially those in upland settings from 6,000 to 8,000 feet in elevation, represent "systemic and repeated long-term use," with this pattern of seasonal movement extending into the Late Archaic period.

A substantial amount of work has been done in the Albuquerque Basin as a result of survey and data recovery projects, and this has greatly enhanced our knowledge of the Archaic period occupation of this area both pre- and post-agriculture (Railey 2016). These projects have documented the consistent presence of Middle Archaic base camps and specialized procurement sites across the landscape, suggesting patterned seasonal land use perhaps tied to the expansion of grasslands, wetlands, and piñon after the Altithermal.

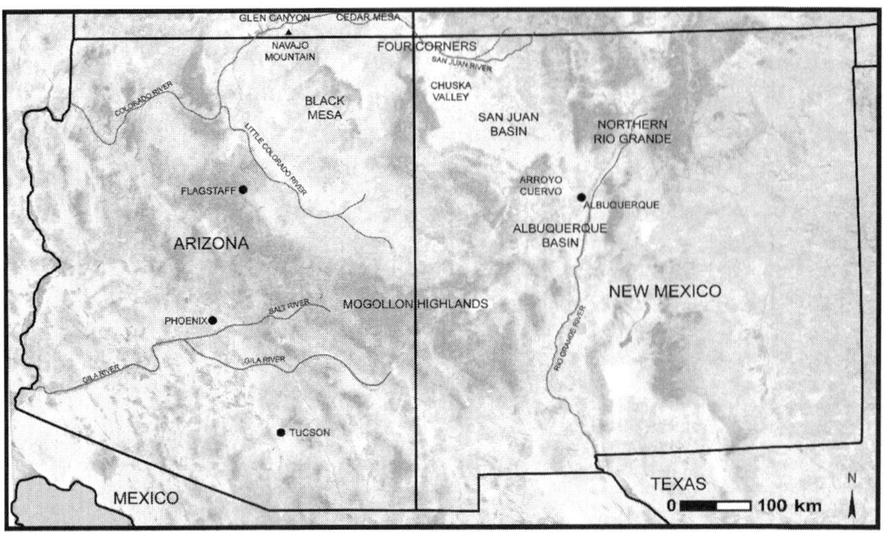

Figure 4.3. Regions of the northern Southwest discussed in this chapter. Figure by Ben Van Alstyne

Agogino and Hester (1956) initially reported on Middle Archaic period sites with hearths, chipped and ground stone, and San Jose points found in dunes located west of Albuquerque. They argued that the predominant cultural influence at these sites was from the south, based on the percentage of ground stone present and the fact that the projectile points resembled Chiricahua points found in the southern Southwest. Their interpretations of a southern origin have not been supported by recent research, however (Vierra et al. 2012), and instead these sites appear to be part of seasonal foraging activities practiced by local hunter-gatherers in the region.

Similar sites were recorded in dunes near the San Jose River outside Grants, New Mexico, by Bryan and Toulouse (1943). These sites contained high artifact densities and abundant faunal remains, and Bryan and Toulouse (1943) suggested that they were repeatedly occupied and represent more intensive occupation than that seen at other Middle Archaic period sites. They suggested the possibility that agriculture may have been practiced by these Middle Archaic groups, but given that the sites were investigated prior to the advent of flotation analysis and no pollen samples were taken, this could not be supported. It is interesting to posit that perhaps today they would be included on the list of early agricultural sites. Further survey in this region has established a substantial Middle Archaic presence, perhaps tied to the presence of grasslands and in some cases, playas.

Several Middle Archaic period occupations in the Rio Grande, Puerco, and Albuquerque regions have yielded structures (Irwin-Williams 1968; Railey 2016; Schmader 2001). These are generally shallow and ephemeral features, and are surprisingly similar to "house pits" identified at Middle Archaic sites in Wyoming that are argued to be associated with short-term occupations by residentially mobile groups (Larson 1997; Smith 2003; Smith and McNees 1999). Smith (2003) argues that they reflect stable land-use patterns. Similar Middle Archaic period structures have been found in the southern Southwest in the Tucson Basin and El Paso regions, as discussed in chapter 5. The use of these structures represents an increase in logistical mobility, probably tied to seasonal reoccupation and, like the Wyoming structures, may indicate a stable land-use pattern.

Moving further south and west, Wendorf and Thomas (1951) defined the Concho Complex based on Archaic period lithic assemblages from sites in east-central Arizona. The sites were found in grasslands near lakes, streams, and springs, and the assemblages indicate a focus on generalized hunting and gathering activities. The Concho Complex may be one component of a seasonal pattern of occupation rather than a separate complex, however. The projectile points resemble San Jose/Pinto points (Wendorf and Thomas

1951:112) and may represent regional variation as Middle Archaic groups adapted to different environmental circumstances. This interpretation is supported by data from the Little Ortega Lake and Laguna Salada sites reported by Martin and Rinaldo (1960), as these are camp sites located on playa beaches with both Pinto and San Jose projectile points and abundant ground stone and show a focus on plant resources along the playa margins.

Work in the Chuska Valley and the Four Corners region of the San Juan Basin has increased our knowledge of the Archaic period presence due in large part to data recovery projects (figure 4.3) (Del Bene and Branchard 1994; Kearns in press; Kearns and McVickar 2007; Murrell and Vierra 2014). Like those occupations described in other regions, there are a number of Middle Archaic camps sites that have been investigated both in rock shelters and open-air sites; these, too, are characterized by seasonal mobility and generalized hunting and gathering activities.

In the San Juan Basin of southwestern Colorado and northern New Mexico, Kearns (in press) notes an increase in site density during the Middle Archaic period that continued into the Late Archaic period. Like those in adjacent areas, these sites represent a subsistence-settlement strategy of seasonal movement, with some evidence of longer periods of occupation at sites with structures (generally small oval, shallow structures). Bell-shaped storage pits were found at one site, LA 80392, near a playa in the southern Chuska Valley dating to 3000 BC (Kearns in press). The presence of bell-shaped storage pits at this site is unique, as in other portions of the Southwest these features co-occur with the adoption of maize (see chapter 2). Kearns (in press) suggests that these pits were used to cache surplus seeds collected during the summer for later retrieval, following a strategy first suggested by Elyea and Hogan (1993).

The common presence of Middle Archaic period sites and components throughout much of the northern Southwest contrasts substantially with data from portions of the northern Colorado Plateau in southeastern Utah and northern Arizona, especially in areas where extensive archaeological work has been done such as Glen Canyon, Cedar Mesa, Black Mesa, and the Navajo Mountain area (figure 4.3). Schroedl (1976) was the first to remark on the lack of evidence of Archaic period occupations in this portion of the northern Colorado Plateau, and recent projects have confirmed a lack of both Middle and Late Archaic period sites in the region (Matson 1991; Parry et al. 1994; Powell and Smiley 2002; Geib 2011). For example, Matson (1991:10) reports that only five Archaic sites were recorded on Cedar Mesa, most of which were temporary camps, leading him to declare that the Basketmaker II "appeared out of nowhere."

This lack of Archaic sites could be the result of recovery bias, as Geib (2011) has noted that many of the Archaic sites found during work on the Navajo Mountain Road project in northern Arizona were not visible on the surface. It is also possible that the environment played a major role in the low number of Archaic period sites and the sporadic use of this region until post-Archaic times. The environment is characterized primarily by shrub grassland, with some grasses such as dropseed and Indian ricegrass, and piñon is present in some upland settings. It is generally poor hunting habitat. Lenses of sterile soil at Dust Devil and Cowboy caves indicate that the Middle Holocene was extremely dry in this region (Geib 1996); thus, an area already lacking in reliable water and resources may have become even less attractive to foraging groups. Although the environment in this area is not highly productive, it is surrounded by areas that are, and these have substantial Archaic period occupations. It is thus possible that Archaic foragers did not consider this portion of the northern Colorado Plateau to be suitable for occupation until agricultural production could sustain a population. The initial farming adaptation in this region would then be Basketmaker II and not Archaic, a point discussed in more detail below.

Agricultural Beginnings

There are currently six sites with evidence of early maize in the northern Southwest; three open-air sites, one rock shelter, and a packrat midden with maize pollen (table 4.1; figure 4.4). Simmons (1982, 1986) was one of the first to find evidence for the use of early maize by Archaic groups in this area. He recorded a number of small artifact scatters in the San Juan Basin located along Gallo and Alemita washes east of Chaco Canyon in northern New Mexico at an elevation of 6,200 to 6,400 feet. Several of these were

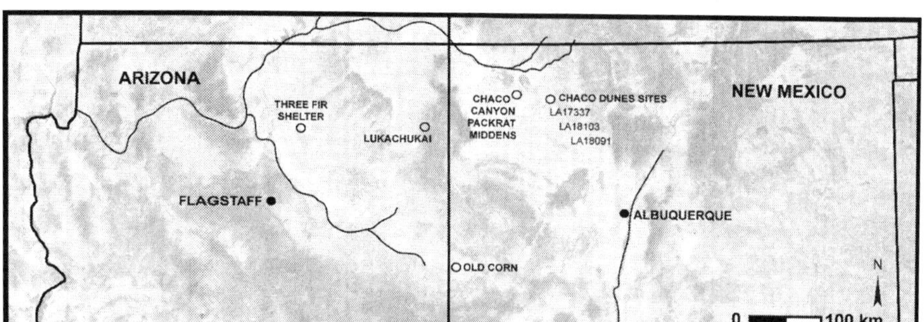

Figure 4.4. Sites with early maize in the northern Southwest. Figure by Aaron Woods

Table 4.1. Sites with Earliest Maize Dates in the Northern Southwest

Site	*Dates (Radiocarbon Years BP)*	*Calibrated BC Dates/ Intercept*	*Reference*
Old Corn, New Mexico	3810 ± 50	2460–2060 BC/2200	Huber 2005
	3760 ± 50	2350–2030 BC/2180	
	3730 ± 40	2280–1990 BC/2130	
	3710 ± 50	2280–1960 BC/2100	
	3700 ± 50	2270–1950 BC/2090	
	3680 ± 40	2200–1950 BC/2070	
	3660 ± 50	2200–1900 BC/2040	
	3660 ± 40	2190–1930 BC/2040	
	3640 ± 40	2140–1900 BC/2010	
	3620 ± 40	2140–1880 BC/1980	
Chaco Packrat Midden*	3890 ± 40	2490–2260 BC/2370	Hall 2010
	3030 ± 50	1320–1160 BC/1290	
Bat Cave	3740 ± 70**	2430–1950 BC/2150	Wills 1988a
Three Fir Shelter	3610 ± 170	2200–1740 BC/1990	
LA 17337 (Chaco Dunes)*	3680 ± 85	2145–1910 BC/2080	Simmons 1986
	3560 ± 95	1930–1735 BC/1910	
LA 18103 (Chaco Dunes)*	3985 ± 155	2690–2230 BC/2510	Simmons 1986
	3650 ± 70	2080–1890 BC/2035	
Lukachukai	3455 ± 45	1890–1640 BC/1780	Gilpin 1994
	3135 ± 45	1490–1325BC/1410	

*Dates on charcoal associated with maize pollen
**Excavator questions reliability of maize date

small seasonally occupied camp sites located in dunes. The projectile points recovered from these sites were classified as San Jose points, placing them within the Middle Archaic period. Flotation and pollen remains indicated primary reliance on Indian ricegrass and chenopodium at these sites, and Simmons (1982:173) argued that Archaic period foragers were drawn to this area because of the diversity of vegetation available in the dunes.

Four of the sites, three open-air dune sites (LA 17337, 18103, 18091) and one rock shelter (Sheep Camp Rock shelter) yielded maize pollen and/or charred macrobotanical remains indicating that maize was used to supplement wild foods (table 4.1). Five hearths were excavated at LA 17337, a lithic scatter on a dune near Gallo Wash. One hearth yielded maize pollen associated with radiocarbon dates on wood charcoal of 2000–1900 BC (figure 4.5). Maize pollen was also recovered from several of the other hearths, along with cheno-ams (chenopodium-amaranth) and dropseed. Maize pollen was recovered from a rock-lined hearth associated with wood charcoal dating to 2500–2000 BC at a second small lithic scatter, LA 18103. Excavations at a third lithic scatter, LA 198091, found

Figure 4.5. Hearth with maize pollen, LA 17337. Photo courtesy of Alan Simmons

on a dune near Alemita Wash yielded thirteen features, primarily pits. Three bell-shaped storage pits were found, two of which contained charred maize. One of these yielded a radiocarbon date on maize of 890 BC, while wood charcoal from the other pits with maize dated to 1300–800 BC (table 4.1). This site has been interpreted by Simmons (1982) as a seasonal campsite located near agricultural fields and used for processing maize.

In the 1980s, these sites represented the earliest maize found in the Southwest at the time and were not immediately accepted, as discussed in chapter 2. These remains continue to challenge existing views that maize resulted in substantial changes in hunter-gatherer lifeways, as that is clearly not the case with these finds. No evidence of sedentism or an agricultural focus was found at any of these sites and Simmons (1982) argues that the maize represents "experimentation with a new resource by indigenous groups." He proposed that maize was planted in the summer in the open grasslands and harvested in the fall before groups moved to winter rock shelters. He thus sees the use of cultigens as occurring early, but argues that dependence developed much later (Simmons 1986:74). Hogan (1994) has suggested that these sites represent a strategy of planting crops near summer gathering areas, with groups maintaining fields during forays from the uplands to the lowlands. Harvested crops were stored in bell-shaped pits for later retrieval, with small camps used to exploit stored maize.

A final site excavated during this project, Sheep Camp Shelter (LA 29SJ178), yielded early squash remains. The site is located above the floodplain of Gallo Wash at an elevation of 6,500 feet. Excavated deposits yielded evidence of "several ephemeral occupations dating from the Archaic and Anasazi periods" (Simmons 1986:77). Two carbonized domesticated squash seeds were recovered from the Archaic period deposits and one of these dated to 1100 BC (the second dated to 400–300 BC) (table 4.2). This is the earliest documented evidence of squash remains in the Southwest (Simmons 1986:77), along with a squash seed from Bat Cave. Carbonized maize was also recovered from the shelter dating to 400–200 BC, and other plant

Table 4.2. Sites with Earliest Dates on Squash and Beans in the Southwest

Site	*Dates (Radiocarbon Years BP)*	*Calibrated BC Dates/ Intercept*	*Reference*
Squash			
Sheep Camp Shelter	2900 ± 230	1330–810 BC/1122	Simmons 1986
Bat Cave	2980 ± 120	1310–990 BC/1200	Wills 1988a
Tularosa Cave	1900 ± 70	AD 75–245/AD 110	Wills 1988a
Beans			
Bat Cave	2140 ± 110	280–4 BC/190	Wills 1988a
Tularosa Cave	2470 ± 250	820–220 BC/570	Wills 1988a
Fresnal Shelter	2085 ± 60	150 BC–AD 10/110 BC	Tagg 1996
	2015 ± 65	90 BC–AD 90/30 BC	

remains from these deposits included weedy annuals. As with the open-air dune sites, Simmons (1986) interprets this as representing a seasonal campsite with maize used as a supplemental resource.

Two finds support Simmons's data. Hall (2010) reported on maize pollen grains recovered from a packrat midden near Casa Chiquita within Chaco Canyon that dated to 2400 BC, with a second sample dating to 1300 BC (table 4.1). Hall argues that the maize was grown nearby, likely along Chaco Wash, as pollen does not travel far, although it could have been carried by packrats who range more widely. If it does represent early agriculture, then its presence in a canyon bottom suggests that early maize farming was done in alluvial settings where grasses were abundant. Hall notes that this may have been a marshy environment that would have been ideal for both growing maize and for gathering efforts. Unfortunately no archaeological sites with early maize have been identified in Chaco Canyon, although Archaic sites have been recorded in the area and Hall (2010) suggests that the farmers were living at Atl Atal Cave, an Archaic period site located on a nearby mesa top. If this is the case, then as with Simmons's data and the Old Corn site discussed below, the earliest maize in this portion of the Southwest was associated with logistical foraging activities.

The Old Corn site (LA 137258) is an open-air multicomponent site located adjacent to Carrizo Wash, a portion of the Little Colorado Watershed, in west-central New Mexico (figure 4.4; table 4.1) (Miljour and Huber 2005). It sits at an elevation of 6,200 feet at the base of a sandstone mesa that has juniper-covered slopes, with cacti, chenopodium, amaranth, and grasses on the site. Thirty-four features were excavated, including hearths, roasting pits, small basin pits, and bell-shaped storage pits (Miljour and Huber 2005). Thirteen of the excavated features contained maize, with dates clustering in the 2100 BC range (2200–1900 BC) (table 4.1). This is currently the oldest recorded site with maize remains in the northern Southwest, although the dates closely parallel those from other early sites discussed here and the Los Pozos, Las Capas, Clearwater, and Montoya sites in the southern Southwest (see chapter 5).

Huber (2005) interprets the occupation at the Old Corn site as a seasonal camp/processing site occupied during the late spring and summer. In addition to maize, chenopodium and dropseed were found in flotation samples; the lack of piñon indicates that the site was not occupied into the fall. Maize was apparently grown adjacent to the wash using floodwater techniques. Ground stone was present in the assemblage, but Huber (2005) notes that investment in grinding technology was low; this parallels what Simmons (1982) found at the Chaco Archaic sites. Faunal remains indicate a focus on small game, primarily rabbits. No projectile points were found associated with the Archaic period component, but the dates place its occupation in the Middle Archaic period. Huber (2005) sees this site as representing a component of a logistical subsistence-settlement strategy with maize used as a dietary supplement.

Early maize has also been recovered from the Lukachukai site (AZ:I:39:53), an open-air site located near Chinle Wash in the Chinle Valley of northeastern Arizona at an elevation of 6,400 feet (figure 4.6). Only a small portion of the site was exposed during trenching for a waterline, but three small shallow basin-shaped pithouses, four hearths, and one bell-shaped storage pit were excavated (Gilpin 1994). These were associated with maize dated to 1700 BC from the pithouse and 1400–1300 BC from the hearths (table 4.1). Although slightly later than the dates from the Chaco region and the Old Corn site, the early maize is associated with a similar pattern of seasonal land use, although the presence of structures indicates the occupation was less ephemeral than elsewhere. Like the sites in the Mogollon Highlands discussed by Wills (1988a), this site appears to have been occupied during the growing season, thus representing one component of a seasonal round.

Two early maize dates have been reported from rock shelters in the northern Southwest, one from Three Fir Shelter on the Colorado Plateau and one from Bat Cave in the Mogollon Highlands (figure 4.4; table 4.1). Wills (1988a) noted that the sample from Bat Cave, which dated to 2100 BC, might have been contaminated during curation, so it is not discussed further here. Maize dating to 2000 BC has been recovered at Three Fir Rock Shelter located in the piñon-juniper zone on the northern end of Black Mesa in northeastern Arizona at an elevation of 6,500–6,600 feet

Figure 4.6. Lukachukai site (AZ I:39:53). Photo courtesy of Dennis Gilpin

(Smiley 1994). The context of this early maize date is not clear, however, so even if the date is accurate, it is not possible to determine the nature of the occupation associated with early maize use. There is a substantial occupation at Three Fir Shelter associated with maize dating after 1000 BC, and this has been the focus of investigations at the site (Smiley 1994). The site was repeatedly occupied into the Basketmaker II period, with storage pits, slab-lined storage cists, roasting pits, hearths, and cached maize found in deposits dating from 800 BC–AD 100. Maize use at this site may be tied to an overwintering strategy and the evidence of repeated use points to long-term, relatively stable use of this rock shelter. If the early maize date is accurate, then this stable land-use strategy may extend back another one thousand years.

The recovery of the earliest maize primarily at open-air sites indicates that the interpretation put forth by Smiley (1994) that rock shelters in the northern Southwest were the primary locus of occupation for early farmers in the region may be more complicated than he originally posited. It appears that on the southern Colorado Plateau both open-air and rock shelters were used by early farmer-foragers, but it is possible that as you move farther north, rock shelters were used primarily for overwintering and thus reflect a specific pattern associated with maize use. This pattern appears to postdate the initial use of maize, but more well-dated sites with early maize are needed to fully evaluate this.

The open-air sites with the earliest maize discussed here are all seasonal camps and, according to their excavators, do not exhibit either high agricultural dependence, nor are they associated with sedentary occupations. The sites share certain traits, however, particularly the use of bell-shaped storage pits. Bell-shaped pits also appear at the earliest agricultural sites in the southern Southwest and are likely one of the technologies introduced with early maize. The resulting ability to store both maize and wild resources may have been an impetus for the changes that we see during the succeeding Late Archaic period, including increased efforts focused on storage. All of these earliest open-air sites also exhibit high incidences of cheno-ams, perhaps associated with agricultural fields. If these open-air sites are part of a seasonal round that included the incorporation of maize, then more information from surrounding sites that served as part of this seasonal pattern of movement will be useful in establishing the nature of the integration of maize agriculture into the diet.

Lukachukai yielded untyped corner-notched projectile points, the two sites in the Chaco region with maize pollen yielded San Jose points, and the Old Corn site did not have any projectile points (table 4.1). The similarities in overall adaptation represented by these early sites contrast with the lack of consistency in projectile points and further support the idea that projectile points may not be the best indicators of interaction and movement during the Archaic period (Roth 2014).

Summary of Agricultural Beginnings

The northern Southwest thus appears to be characterized by a distinct pattern of agricultural beginnings. Current data indicate that cultigens were adopted by seasonally mobile hunter-gatherers by at least 2100 BC and that maize agriculture initially had little impact on these groups. Maize was integrated into the existing seasonal foraging system, suggesting that it was adopted because it fit into the exploitation of other subsistence resources. This is consistent with ethnographic hunter-gatherer behavior, as discussed in chapter 3.

Vierra and McBrinn (2016) have suggested that in the northern Rio Grande area, groups practiced a system of "plant and go" where they would clear fields and plant in the spring after the last frost and return in late summer or early fall to harvest the crop. This strategy has been referred to as "ancillary" cultivation by Freeman (2012), who argues, using ethnographic data, that these kinds of strategies occur in areas with high wild-resource abundance and are associated with low levels of storage and social networks that enable access to seeds if crops fail. His survey of ethnographic groups indicated that this strategy usually occurred with low population densities and was characterized by subsistence flexibility.

In contrast, Wills (1992) has proposed that initial maize use in the Mogollon Highlands involved some labor investment to tend and monitor the fields in order to ensure crop success. This strategy, termed "surplus cultivation" by Freeman (2012), is associated ethnographically with groups who used cultigens to counteract periodic food shortages. These groups generally have higher population densities than those practicing ancillary strategies and exhibit higher levels of storage.

Hogan (1994) has argued that scheduling conflicts would occur for early farmers in the northern San Juan Basin because maize required labor investment during germination and harvesting, and both were periods of peak wild plant availability. He proposes that crops were planted and then fields were maintained during foraging forays from the uplands to lowlands, with harvested maize cached in dunes for later retrieval.

These do not necessarily have to be opposing views, as different strategies may have been used in different areas depending on wild resource availability, existing foraging strategies, and the environmental setting. Vierra and McBrinn (2016) allude to this in mentioning that the Mogollon Highland sites discussed by Wills (1988a,b) and Wills and Huckell (1994) exhibit more evidence of logistical mobility strategies, which may have favored a degree of labor investment in maize cultivation, and this led to the surplus strategy described by Freeman (2012).

Irrespective of the strategies involved, most researchers working in the northern Southwest see the initial use of maize as fitting into existing foraging regimes, with efforts to grow maize tied to the increased predictability and reliability made possible through cultivation, and the availability of wild plant resources to buffer crop failure. This warranted the efforts placed into cultivation, which fits with how hunter-gatherers would incorporate a new food into their diet (see chapter 3). The risks associated with cultivation in these circumstances would be mitigated by the enhanced foraging possibilities (Wills and Huckell 1994), by ensuring access to needed nutrients, and by attracting other resources such as game, birds, and edible weeds. The storability of maize, and especially its ability to be used as a component of an overwintering strategy, may have been one of the major reasons for its spread across the northern Southwest, particularly to the Colorado Plateau.

The Late Archaic Period

The period postdating the initial adoption of agriculture in the northern Southwest (1200 BC–AD 200) has variably been referred to as the Late Archaic (Vierra 2008; Vierra and Ford 2006), Basketmaker (Smiley 2002), Early Agricultural (Geib 1996), Archaic-Formative (Geib 2011), or Pre-Pottery Formative (Matson 2006) period. I have opted to refer to it as the Late Archaic period because it directly postdates the Middle Archaic period, and I follow Geib (1996, 2011) and Matson (1991) in considering Basketmaker as associated with a discrete set of features and material culture that distinguish it from Archaic occupations.

In this section, data are presented from each of the areas of the northern Southwest where we have solid evidence of what happened after maize was initially introduced, although these data are variable, and there are significant gaps in our knowledge of some areas, such as the Lower Colorado River region. The differing interpretations of these occupations are also presented, and it should be apparent from this discussion that differences occurred in the way that maize was integrated and its subsequent impact on Ceramic period developments.

Mogollon Highlands and Upper Little Colorado River

The Late Archaic occupation of the Mogollon Highlands is perhaps the best known one because maize from Bat Cave was initially used as the argument for a very early entry of agriculture into the Southwest (see chapter 2). A substantial Late Archaic presence has been documented in this region and has been most extensively discussed by Wills (1988a,b, 1989, 1992, 1995). These occupations have generally been classified as San Pedro Cochise because of the presence of San Pedro style projectile points; however, they are included in the discussion of the northern Southwest here because of their location in upland settings versus the lowland desert settings characteristic of the southern Southwest.

Bat Cave represents one of the earliest sites with maize in the Mogollon Highlands, with maize remains dating to 1000 BC (figure 4.7; table 4.3). The cave actually consists of a number of adjacent rock shelters, excavated first by Dick in the 1950s (Dick 1965) and later by Wills in the 1980s. Wills's (1988a) excavations documented multiple episodes of occupation represented by occupation surfaces with hearths and pits. In addition to maize, evidence for the exploitation of a wide variety of wild plants was found, including piñon, walnut, acorns, banana yucca, prickly pear, and cheno-ams. Cattail was found in one pit, indicating the presence of standing water

Table 4.3. Late Archaic Sites with Maize in the Northern Southwest (before 800 BC)

Site	Dates (Radiocarbon Years BP)	Calibrated BC Dates/ Intercept	Reference
Bat Cave	3120 ± 70	1480–1315 BC/1380	Wills 1988a
	3060 ± 110	1430–1130 BC/1285	
	3010 ± 150	1430–1005 BC/1220	
Jemez Cave	2900 ± 40	1335–1000 BC/1250	Ford 2013
San Luis de Cabezon	3125 ± 45	1395–1280 BC/1340	Railey 2016
	3055 ± 45	1335–1210 BC/1320	
Salina Springs	2630 ± 40	850–550 BC/810	Gilpin 1994
LA 01891	2720 ± 265 (maize)	1180–500 BC/890	Simmons 1986

nearby, and Wills (1988a:100) suggests that there may have been water in nearby playas. Wills (1988a, 1992) proposes that the way the site was used changed over time, with a shift to a logistical strategy during the Late Archaic following the adoption of maize at 1000 BC, characterized by occupation floors with storage pits and an increase in artifact density and diversity.

Evidence from Tularosa Cave also points to the use of upper elevation sites (6,500–6,700 feet) as base camps during the Late Archaic period in the Mogollon Highlands (figure 4.7; Martin et al. 1952; Wills 1988a).

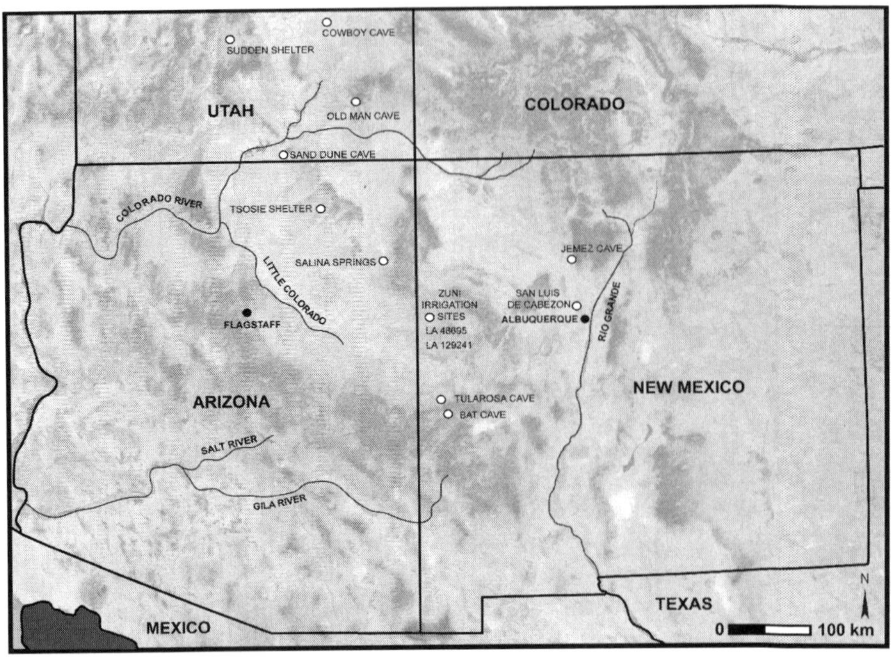

Figure 4.7. Late Archaic sites in the northern Southwest. Figure by Aaron Woods

The site yielded abundant maize along with large storage pits, postholes that may represent structures, and a large number of perishables (baskets, sandals, rope, etc.). Radiocarbon dates place the occupation at 300 BC, with dates on maize of 100 BC (table 4.3), and the cave was occupied into the initial Ceramic period. Numerous snares were found in the Archaic period levels; Wills (1988a:131) notes that hunting was more extensive at Tularosa Cave than at Bat Cave. A squash seed recovered from the lowest levels of the cave dated to 100 BC, and a direct date on a bean from the cave dated to 600–300 BC (table 4.2), representing the earliest bean remains in the Southwest. Nearby Cordova Cave also yielded evidence of maize associated with the Late Archaic occupation dating to circa 300 BC, but no storage pits or structures were found. Wills (1988a) argues that this was a hunting camp with limited maize use.

Tularosa Cave exhibits evidence of continuity from the Late Archaic to the earliest Mogollon, based on the stratigraphy and continuity in material culture. Thus, the questions raised by research in this area differ from those faced by researchers in other portions of the Southwest where evidence for continuity is lacking, such as the Colorado Plateau. The focus of research in the Mogollon Highlands has been on how and why maize was integrated into foraging adaptations and the subsequent impact of the intensification of maize agriculture and the adoption of squash and beans. Wills and Huckell (1994) have used data from these sites in constructing their argument that the initial use of maize was linked to efforts to enhance foraging efficiency. They argue that maize would allow groups to move to the uplands earlier than previously possible, and thus they were able to monitor wild resource availability. This would have helped with their foraging efforts, especially in light of competition from other foraging groups exploiting the uplands. Wills (1995) sees the interplay between maize farming, increasing population densities, and the resulting competition for high-ranking resources from lowland groups as responsible for the continued focus on maize agriculture in this area.

Northern Rio Grande Valley

The earliest evidence of maize in the northern Rio Grande area comes from Jemez Cave, where maize dating to 1250 BC has been found (figure 4.7; table 4.3; Ford 2013). The cave contained evidence of a Late Archaic occupation that Ford ties to the presence of a seasonal lake below the site and the diversity of wild resources available within short distances. Ford (2013) sees relatively episodic occupation of the cave, as occupation layers were separated by sterile lenses, suggesting occupational hiatuses that Ford thinks could have lasted a decade or longer. He argues that the cave was occupied during the spring and then again during the late summer and fall. Domesticated squash remains were recovered in the Late Archaic period levels, which may represent early squash like that recorded at Tularosa Cave and Sheep Camp Shelter, but no dates are available on them.

Vierra (2008) has used data from Jemez Cave and other sites in the northern Rio Grande region to argue that Late Archaic period hunter-gatherers integrated maize into their seasonal subsistence strategies. The earliest sites with maize are all located in well-watered settings near piñon-juniper stands, a pattern similar to that documented in the Mogollon Highlands. Vierra and Foxx (2010) and Post (2013) have proposed that Late Archaic groups in this region practiced a seasonal land-use strategy that began during the Middle Archaic period, with groups going from lowlands in early summer where they exploited ricegrass, then moving to upland Ponderosa pine forests in midsummer where they exploited cheno-ams and hunted deer, then down to the piñon-juniper zone in late summer and early fall to exploit piñon, juniper berries, acorns, and game. Groups would then have winter camps along the lowland rivers, where several sites have yielded Late Archaic period structures (Post 2013).

The Late Archaic occupation of the Rio Grande region exemplifies the flexibility of hunter-gatherer responses to the introduction of agriculture, as foragers in this region sometimes, but not always, incorporated maize agriculture into their subsistence regime. Some Late Archaic sites have structures and exhibit evidence of logistical land-use strategies but lack maize, instead revealing a focus on wild plants and hunting. Vierra and Ford (2006:501) see this as illustrating that maize was "differentially integrated into local foraging economies." They argue that the primary focus remained on wild resources, especially cheno-ams in the lowlands and piñon and game in the uplands, and maize represented a resource that could be used as a backup in case of resource shortfalls.

This stable foraging strategy continued in the Rio Grande region well into periods when maize use intensified in other areas, and it was not until about AD 600 that evidence of intensification is apparent in changes in site structure indicating increased sedentism, more emphasis on storage, and higher population densities (Post 2013; Vierra and McBrinn 2016).

Albuquerque Basin

Railey (2016) has summarized the growing database on the Late Archaic occupation of the Albuquerque Basin as a result of cultural resource management projects. He notes that a large number of Late Archaic sites have structures, which are generally small, shallow, circular or oval pits similar to Middle Archaic period structures and that these represent a shift to more logistically organized settlement and subsistence strategies. Ten sites have yielded maize, and all of these are located in the river valleys, indicating that maize was first adopted in productive alluvial settings similar to those found in the southern Southwest.

The earliest maize in this area has been recovered from the San Luis de Cabezon site (LA 110946) on the Rio Puerco floodplain northwest of Albuquerque (figure 4.7; table 4.3; Bargman et al. 1999; Railey 2016).[1] Seven structures were excavated, and associated maize remains dated to 1400–1200 BC. An earlier, pre-maize occupation is also present at the site, with shallow oval houses and evidence of reliance on a variety of grasses and cheno-ams, suggesting that maize was added to a preexisting foraging strategy of seasonally intensive occupation of the floodplain focused on wild plant gathering, akin to what Roth and Freeman (2008) found on the Santa Cruz River floodplain in the southern Southwest. Vierra and McBrinn (2016) note that maize at the Cabezon site is "more developed" than maize at Jemez Cave, which exhibited more primitive characteristics and was consequently less productive (Ford 2013). They see this as indicating more selection and tending of the lowland valley maize than at upland sites.

In contrast to the Cabezon site, Schmader (2001) reports on sixteen Late Archaic period structures excavated in the Albuquerque region that did not yield maize remains but instead yielded a substantial quantity of cheno-ams, further documenting the variability in the incorporation of maize by Late Archaic foragers in the region. Bison bones have been found at several Late Archaic sites in this region, and hunting still formed a major dietary component, perhaps accounting for the lack of emphasis on cultigens despite their known presence.

Railey (2016) argues that maize was added as part of a diversification strategy to increase carrying capacity as Late Archaic groups began to occupy prime farmland and territorial circumscription developed. Farming dependence increased by the En Medio phase (800 BC–AD 400) in the Albuquerque Basin, a pattern consistent with that observed in other areas. Railey notes that some groups, such as those on the West Mesa outside Albuquerque, may have remained hunter-gatherers while others adopted cultigens. It is this kind of complex interaction between farmers and foragers during the Late Archaic period as cultigens were adopted by more groups that necessitates moving beyond simply documenting the earliest maize to explore the social dynamics that developed as some groups became farmers and others remained foragers.

Northwestern New Mexico

One of the most significant finds concerning early agriculture in the northern Southwest comes from northwestern New Mexico, where evidence of irrigation has been found along the Zuni River on the Colorado Plateau dating to 1000 BC (Damp et al. 2002). Canals were found at K'yana Chabina (LA 48695) and K'yawa:na'a (LA 129241), and maize pollen recovered from them confirms their use for farming. The features include both runoff-supplied canals from a tributary wash of the Zuni River and canals located on the main floodplain of the river used to divert river water. They are significant because they represent the first evidence of agricultural intensification in this portion of the northern Southwest. Damp et al. (2002) argue that they represent the spread of irrigation technology from southern Arizona and northern Mexico; this is likely, given the presence of canals in the Tucson Basin by 1500 BC and the potential for contact between farmer-foragers in these areas, although it is also possible that they represent the movement of farmers into the well-watered Zuni region (Damp 2007). More information is needed on the Archaic period occupants of this region before this can be determined. The benefits of irrigation in this region would have been substantial, as it would allow crops to be planted early, right after the last frost, and would have increased the productivity of maize fields.

San Juan Basin

Much of the data on the Late Archaic occupation of the San Juan Basin comes from cultural resource management projects, several of which were large-scale surveys. These projects have helped place the Late Archaic period in regional context and provide important insights into how maize agriculture spread and the varying consequences of its introduction across this region, as many of the recorded and excavated sites represent the spread of maize to hunter-gatherer groups after the successful integration of maize agriculture at sites like Lukachukai.

Early maize use at the Lukachukai site in northwestern New Mexico was followed by what appears to be an increased focus on maize agriculture, with consequent changes in settlement strategies associated with increased sedentism. The Salina Springs site (AZ:I:63:7) is located southwest of the Lukachukai site in a piñon-juniper woodland at an elevation of 6,620 feet (figure 4.7; Gilpin 1994). The site has Basketmaker III and pueblo-period components, but several clustered features dating to the terminal Late Archaic period were also found, including one pithouse.

Maize dates place its occupation at 800–400 BC (table 4.3). The pithouse was larger than the shallow oval pit structures described in the area thus far, likely tied to increased sedentism (Gilpin 1994:240).

Evidence of maize agriculture has also been found in the Chuska Mountains in the western San Juan Basin to the east of the Lukachukai site. The earliest maize in this area was found at LA 6444, a site with numerous structures representing multiple seasonal occupations. One structure yielded maize dating to 788 BC, and as with many of the sites in the area, cheno-ams and ricegrass were also found (Vierra 2008).

Toll and Cully (1994) remarked on the prevalence of Middle and Late Archaic period sites in dunes in the Chuska region, with occupations generally focused on the exploitation of spring and summer annuals including ricegrass, dropseed, cheno-ams, and maize. They argue that these sites were part of an overall settlement system that involved seasonal use of lowland settings during the spring and summer with movement to uplands such as the Chuska foothills in the fall.

Part of this seasonal strategy of maize use during the Late Archaic period may have involved field caching. Work at the Dog Leg site (LA 6448) located near a playa in the San Juan Basin revealed forty-one storage pits with abundant maize dating to 300–100 BC (Kearns et al. 1998). The excavators interpret this as a field cache as, despite the large number of storage features, there were no habitation structures present. Twelve burials were found, and Kearns et al. (1998) suggest that these represent the beginning of marking land tenure, documenting a shift in the way that land was perceived and used and indicating that arable land near water sources was becoming important and perhaps restricted, similar to what Railey (2016) found in the Albuquerque Basin.

Late Archaic period sites have also been recorded throughout the northern San Juan Basin (Mesa Verde region) and appear to represent a pattern similar to that documented in the Rio Grande Valley, with seasonal movement between lowland valleys and upland piñon-juniper and Ponderosa pine zones (Kearns in press). The earliest known site with maize in southwestern Colorado comes from 5LP175, located in the Ridges Basin which yielded maize pollen and cheno-ams dating to 1000–800 BC (Potter 2010a). The excavators interpret this as a seasonal campsite, and thus, as with the early maize remains documented in the Chaco and Rio Grande regions, early maize use was associated with seasonal mobility. This is the only known site with maize remains dating this early in the northern San Juan area, however, and it appears that maize did not become a regular component of subsistence practices in this portion of the Southwest until around 500 BC, indicating a lag of several hundred years before maize was integrated into the subsistence system.

After 500 BC, relatively rapid cultural changes occurred, continuing into the Basketmaker II period. All of the locations where maize remains have been recovered are well-watered settings where floodplain and runoff farming could be done (Kearns in press). Cheno-ams and ricegrass were also common remains found at these sites, indicating that wild resources still formed an important component of the resource base. Several Late Archaic period sites have yielded structures, indicating a logistical mobility strategy that may have facilitated the incorporation of maize agriculture in this area.

Subsequent to the Late Archaic period, a local variant of the Basketmaker II, referred to as the Durango Basketmaker, appeared dating to AD 150–350 (based on tree-ring dates) (Morris and Burgh 1954). Morris and Burgh (1954) reported on excavations of a number of Basketmaker sites near Durango, Colorado, that contained circular structures with evidence of cribbed log superstructures, slab-lined storage cists, and perishables including distinct basketry (Matson 1991). Varien (1999) has suggested that the environment in the area, especially early frosts and unpredictable rainfall, precluded heavy reliance on maize until it was a well-established subsistence strategy that could be sustained. This appears to account for both the evidence at 500 BC for the increased use of maize and the subsequent Basketmaker II intensification.

Additional information on Archaic and Basketmaker sites in the Durango area has come from the Animas-La Plata project (Potter 2010a,b). The Archaic period occupation of this area was sporadic and seasonal. The Basketmaker II sites dated from AD 300–550 and differ markedly in settlement location from earlier Archaic sites, as they are all located on or near soils that were suitable for farming. Potter (2010b) suggests that the presence of slab-lined and bell-shaped pits at the Basketmaker II sites is tied to winter storage, so overwintering was likely an important reason for doing maize agriculture in this area.

Most researchers have argued that these Eastern Basketmaker II groups represent continuity with earlier Late Archaic foragers who initially adopted maize, given strong continuities in chipped stone technology and textiles between them (Charles and Cole 2006; Matson 1991). This relationship remains an issue that needs to be investigated more thoroughly if we are going to understand the role and impact of early agriculture in this region. Although there is no evidence for migration, changes in subsistence and settlement strategies are apparent over time (Charles and Cole 2006; Eddy 1961, 1966), and continued research can shed light on the processes that led to increased sedentism and agricultural intensification.

Colorado Plateau of Northern Arizona and Southeastern Utah—Black Mesa, Glen Canyon, Cedar Mesa

As with the Middle Archaic, data on the Late Archaic period occupation of the Colorado Plateau of northern Arizona and southeastern Utah are lacking, despite large-scale field projects conducted in the region. In contrast, abundant data are available on the Basketmaker period, which has been one of the main factors leading researchers to posit that agricultural groups migrated into the region after the Archaic period. Berry (1982) was the first to argue that there was no relationship between the Archaic period occupants of the northern Colorado Plateau and Basketmaker II; as discussed in chapter 2, he used the absence of dated Archaic period occupations in the area to argue that San Pedro Cochise farmers from the southern Basin and Range region migrated into the area, an idea later supported by Matson (1999, 2002). Geib (2011; Geib and Spurr 2000) has also discussed the fundamental differences between Archaic period and later Basketmaker II occupations on the Rainbow Plateau of northern Arizona, but does not argue that these were San Pedro–phase Cochise immigrants, instead preferring to refer to them as Basketmaker II without defining their place of origin. Matson (1991) has noted a similar discontinuity in Archaic period and Basketmaker II occupations on Cedar Mesa in southern Utah; as noted earlier, this led him to argue that "the Basketmaker II culture, then, appears as if out of nowhere to begin the Anasazi cycle of occupation and abandonments on Cedar Mesa" (Matson 1991:10).

One factor supporting the lack of association of Archaic and Basketmaker II groups in this area is the discontinuity observed in most of the rock shelter sequences excavated in this region, which contrasts with the continuity documented in other areas (e.g., Tularosa Cave). Sites such as Sudden Shelter, Cowboy Cave, Old Man Cave, Sand Dune Cave, and Dust Devil Cave (figure 4.7) contain evidence of Archaic period use dating back in some cases to the Early Archaic period, but they lack evidence of early maize or continuity in material culture between Archaic and later Basketmaker II occupations, possibly because they are not in areas that would have been conducive to early agriculture (Geib 2011). Most of the caves contain sterile deposits that have been interpreted as occupational hiatuses associated with Middle Holocene drying, although these are not always clearly associated with specific time periods and have not been correlated across occupations (Geib 2011).

Using data from the Glen Canyon region of southern Utah, where some Archaic sites have been found, Geib (1996) argues that hunter-gatherers did not entirely abandon the region during the Altithermal, but instead readjusted their foraging strategies to focus on rivers rather than upland settings and expanded their foraging territories. He sees this as accounting for the gaps in occupation observed at Cowboy, Dust Devil, and Old Man Caves.

In contrast, Smiley (2002) has documented more mesic conditions during the Middle Holocene on Black Mesa, based on excavations from Tsosie Shelter (figure 4.7). This suggests that the Altithermal was variable across the Colorado Plateau, as has been documented in other portions of the Southwest (Huckell 1996a), and may have resulted in forager population redistribution as groups responded to varying climatic conditions. Some of the more arid and environmentally marginal areas such as portions of the northern Colorado Plateau were probably abandoned by Archaic period groups in favor of better watered nearby areas on the southern Colorado Plateau and Four Corners area where we see more substantial Middle and Late Archaic period occupations.

In other portions of the northern Colorado Plateau where Archaic sites are present, there is no current evidence for continuity in the occupations into Basketmaker II times. Instead, the Archaic and Basketmaker II occupations appear to represent distinct groups. Geib and Spurr (2000:240) have noted that major differences in site types exist between the limited Archaic period presence found on the Rainbow Plateau and later Basketmaker occupations, arguing that there is no local precedent for these Basketmaker groups. They see the Basketmaker occupation, which is represented by pithouses, bell-shaped storage pits, and maize agriculture, as coming in as a full-blown Basketmaker adaptation by 300 BC. They caution, however, that local discontinuity may not represent regional discontinuity and recognize that maize may have appeared in this region through different mechanisms.

Basketmaker II

Despite the lack of evidence on the Archaic period occupation of the Colorado Plateau of southeastern Utah and northern Arizona, we have abundant evidence of Basketmaker II populations throughout this region dating back to rock shelter excavations conducted in the early 1900s (Kidder and Guernsey 1919; Guernsey and Kidder 1921; Matson 1991; Reed 2000; Geib and Spurr 2000; Geib 2011; Powell and Smiley 2002). When Kidder (1927) first classified Basketmaker II during his development of the Pecos Classification, he dated it from circa 200 BC to AD 400. Subsequent work has generally supported this time frame, with some refinements. Most researchers working in this area see Basketmaker II as associated with a specific suite of material culture traits, including slab-lined storage cists, sometimes with burials in them; atl atls; coiled basketry; sandals; twined woven bags; bone tools; and evidence of the cultivation of maize and squash (Matson 1991; Geib 2011). In contrast, Smiley (2002) has classified all sites with evidence of agriculture on the Colorado Plateau as Basketmaker II, but because these sites lack the characteristics of Basketmaker II as defined by Kidder, this has not generally been followed by other researchers, who continue to date its inception in the 400–200 BC range (Geib 2011). To further complicate the situation, Lipe (1999) has referred to Basketmaker II in the northern Southwest and San Pedro– and Cienega-phase materials from the southern Southwest as Pre-Pottery Neolithic; Matson (2006) then suggested calling the Basketmaker II the Pre-Pottery Formative. None of these terms have caught on, and most researchers continue to refer to sites with this suite of material traits as Basketmaker II. The plethora of ideas about what to call it illustrates the challenges facing researchers working in the northern Southwest in partitioning the diversity of post-Archaic occupations and adaptations into discrete categories, mirroring some of the problems faced in the southern Southwest in attempting to distinguish the Late Archaic and Early Agricultural periods (see chapter 5).

Basketmaker II was initially defined by Kidder (1927) based on excavations done in rock shelters located in the Marsh Pass region of northern Arizona (Kidder and Guernsey 1919; Guernsey and Kidder 1921). These Basketmaker II remains are now classified as the White Dog phase (Matson 1991), and there are several other Basketmaker II variants that have been recognized and may represent either temporal or regional developments (e.g., the Lolomai phase on Black Mesa, which postdates the White Dog phase, and the Grand Gulch phase on Cedar Mesa, which dates to AD 200–400). Taken together, they are all considered to be part of the Western Basketmakers. These groups are distinct from Basketmaker II groups in the Durango region, the Eastern Basketmaker discussed above, based on a number of traits including basketry and architecture. Although Basketmaker II was first classified by Kidder as *pre-pottery*, discoveries of a relatively crude untempered brown ware at sites in the Navajo Mountain region dating to AD 200 suggest that ceramics were first used during this period (Geib and Spurr 2000).

The differences between Western and Eastern Basketmaker groups and the reasons for these differences continue to be of major concern to Southwestern researchers working in the northern Southwest and have implications for our understanding of agricultural origins. The overall Basketmaker

II adaptation is relatively similar across the Eastern and Western regions, with shared traits including slab-lined storage cists, often found in rock shelters, the use of the atl atl, and maize agriculture, but differences in material culture exist between the two. Matson (2006, 2007) has used differences in architecture, projectile points, and textiles to argue that Western and Eastern Basketmaker groups represent two distinct ethnic groups, lending support to his idea that Western groups developed via migration of San Pedro Cochise groups from the south (see also Berry and Berry 1986), while Eastern Basketmaker groups developed from local hunter-gatherer groups who adopted agriculture. As discussed in chapter 2, the data for the connection between San Pedro–phase groups in the south and Western Basketmaker are questionable, but this does not negate the fact that differences between Western and Eastern Basketmaker exist.

Architectural differences between the two are primarily visible in house construction. Western Basketmaker II houses are circular, often with short slab-lined entryways and hearths that usually have a deflector stone facing the entryway. Eastern Basketmaker II houses are also circular but exhibit distinct cribbed-log construction. Webster and Hays-Gilpin (1994) examined Basketmaker textiles and found that there was a considerable amount of diversity in them—with "technological similarities in textiles, significant technological differences in basketry, and some degree of stylistic unity overall" (Webster and Hays-Gilpin 1994:315). They infer that "patterned differences in basketry and textile technologies signify different cultural origins for Basketmaker II people" (Webster and Hays-Gilpin 1994:319), arguing that Western (White Dog) Basketmaker are more closely affiliated with Mogollon and Cochise groups to the south while Eastern Basketmaker groups are more closely affiliated with Great Basin groups.

A comparison of dental traits from seventeen Eastern and sixty-one Western Basketmaker individuals revealed major distinctions between the two, with "extreme divergence of the Eastern Basketmaker sample" from all other Southwestern groups used in the study noted by the authors (LeBlanc et al. 2007). Mowrer (2006) also identified distinctions in burial orientation and in the presence of single versus multiple burials between Western and Eastern Basketmaker groups.

In contrast to these arguments for dramatic differences between Western and Eastern Basketmaker groups, Charles and Cole (2006) have argued that similarities in projectile points, pipes, ornaments, bone tools, and rock art are evident between the two areas and indicate that a shared cultural system was in place by 400–200 BC. They view these similarities as developing in place as Archaic period hunter-gatherers adopted farming and other technologies, with the Basketmaker becoming a more coherent pattern through time.

These divergent views of Basketmaker, based on similar data sets, exemplify the challenges associated with defining prehistoric identity using material culture. Basketmaker II is not a unitary phenomenon, but what do these differences mean, and do they relate to differences in cultural origins? Recent contributions by researchers working with the concept of technological style may have merit in helping address this issue. Geib's (2002) study of projectile point flaking patterns from Sand Dune Cave on the Rainbow Plateau is an excellent example of this. He found very distinct flaking patterns in Basketmaker and Archaic period projectile-point manufacture, indicating the use of a new technique by Basketmaker groups and supporting the presence of cultural differences between the two. These kinds of studies could significantly enhance our understanding of the relationships between local hunter-gatherer groups and Basketmaker farmers. Bone tools might be particularly useful for this kind of study, as distinct forms such as notched-bone tools and antler wrenches have been found at Basketmaker sites throughout the region, and a comparison of these may provide significant information on regional cultural differences. Rock art similarities, differences, and potential continuity also have strong research potential (Schaafsma 1980; Charles and Cole 2006). Geib (1996) cites differences in rock art styles in support of the idea that the Colorado River served as a boundary between Basketmaker groups on the south and Fremont groups to the north. These kinds of studies and comparisons of Archaic period and Basketmaker groups may provide insights into the relationships between them and help determine how far distinct group identities extended into the past.

The Relationship between Late Archaic and Basketmaker

At present, it appears that Basketmaker groups moved into portions of the northern Colorado Plateau, especially far northern Arizona and southeastern Utah, after developing elsewhere rather than developing from local Archaic period groups. As all of the Basketmaker occupations in this region substantially postdate the earliest agricultural sites in the Southwest and thus occur after agriculture was already well established as a component of the subsistence base of farmer-foragers in adjacent areas, this "migration" can be viewed more as an expansion of early farmers into a region not regularly inhabited by hunter-gatherers than as a true mechanism for the spread of early agriculture. Many researchers have noted that agriculture would have been necessary for any long-term occupation of the northern Colorado Plateau, and this may be precisely why we see agriculture come in so much later than in other parts of the Southwest. The reasons for this movement are interesting in light of the fact that by the time farmers moved into the region, they were already dependent on maize agriculture and thus do not represent a transitional phase from foraging to farming (Matson and Chisholm 1991; Coltrain et al. 2007). This indicates that the shifts in adaptation associated with agricultural beginnings occurred elsewhere and were then modified to suit the conditions on this portion of the Colorado Plateau. One caveat remains, however, as this may simply be a function of the database, and if more evidence of Archaic period occupations is found, then this idea may need to be modified.

Smiley (2002) sees the early Basketmaker II period on Black Mesa as representing a long period of "adaptive stability" following the initial introduction of maize into the region, associated with the occupation of rock shelters, low population levels, and some degree of residential mobility. He ties this to the use of maize as a supplement to counteract the variability in yields of piñon masts. He notes that maize is almost always found with piñon in these settings and sees this as a strategy that was viable over a long period of time, with groups relying on stored maize and piñon and following planned and scheduled moves. Thus, the integration of maize agriculture with foraging may also have been an important strategy allowing Basketmaker groups to successfully occupy this area.

Other portions of the northern Southwest such as the southern Colorado Plateau in areas like Lukachukai may be places to look for the development of the Basketmaker pattern. At Lukachukai and Salina Springs, pithouses and storage pits occur early, and it is possible that sites like this represent the initial development of the Basketmaker. Gilpin (1994) has suggested that agriculture was well established by the time that Salina Springs was occupied; this may indicate that farmers were developing agricultural production techniques and maize strains that were adapted to the requirements of the northern Colorado Plateau environment. Similar sites have been recorded in dunes on Gallegos Wash south of Farmington in northern New Mexico, with shallow oval and circular pithouses (lacking the antechambers that are characteristic of later Basketmaker structures) and bell-shaped storage pits dating to AD 1–200 (Del Bene and Branchard 1994). These may also be part of the initial Basketmaker pattern. As these groups became more agriculturally focused and began to develop settlement strategies tied to this subsistence shift (including prolonged occupation of settings where crops could be grown successfully and increasingly sophisticated storage technology), growing populations and competition for land may have led some groups to move into new areas, which may account for the later movement into places like the Rainbow Plateau and Cedar Mesa. Geib and Spurr (2000:240) alluded to this possibility for the Rainbow Plateau, arguing that Basketmaker groups may have developed further south in the Marsh Pass area and that the initial occupation of the Rainbow Plateau occurred as part of the local expansion of farming groups.

Maize on the Fringes—Recent Work in Utah

The Kanab Basketmaker of southeastern Utah have generally been considered to be part of the Western Basketmaker but do not appear to have developed until around 100 BC. Sites with pithouses, slab-lined pits, and bell-shaped storage pits have been recorded in this area, and researchers have argued for the *in situ* development of these Basketmaker groups from Archaic period predecessors in the region (Charles and Cole 2006). Some of the recorded structures have evidence of having had cribbed-log roofs, generally considered to be an Eastern Basketmaker trait, which has led some researchers to question their origins and relationships (Charles and Cole 2006:185).

A large-scale cultural resource management project done prior to the expansion of a reservoir on Jackson Flats near Kanab will contribute to our understanding of Basketmaker occupations in this region; the final project report is currently in progress (Roberts n.d.). One of the most significant finds from this project in terms of examining agricultural beginnings is the presence of a Late Archaic period site that differs substantially from earlier Middle Archaic period occupations in the area. At the Eagle's Watch site (42-Ka-6165), one pithouse and seven extramural features were excavated, including two bell-shaped storage pits (Roberts n.d.). Maize from the pithouse hearth and one of the bell-shaped storage pits dated to 1100–900 BC. The recovery of maize dating to 1000 BC this far north suggests that early farmers expanded into areas with abundant resources, like Jackson Flats, but apparently ignored more marginal areas. The nature of the features and artifact assemblage at Eagle's Watch is particularly interesting, as the remains resemble those from Lukachukai, although this occupation is associated with an Elko projectile point rather than San Jose points. There is a hiatus in the occupation of this area after 800 BC until 100 BC, so the relationship between these early farmers and later Basketmaker groups is currently not known.

Summary

Data on agricultural beginnings in the northern Southwest present a very different picture from that seen in later post-introduction time periods. A growing body of evidence indicates that maize was initially introduced into this region via the adoption of cultigens and storage technology by indigenous Archaic period foragers as a dietary supplement, resulting in a pattern of "low-level food production" (Smith 2001b) characterized by the low dependence and low impact posited by Irwin-Williams (1973) and Minnis (1992). An increasing number of sites with maize dating in the 2000 BC range indicate that initially maize was integrated into existing seasonal foraging strategies at both open-air and rock-shelter sites, which possibly represent different components of seasonal subsistence-settlement strategies. Maize was likely adopted from Mexico through interaction with farmer-forager groups who had already adopted it (Merrill et al. 2009), although the coincidence of dates for the earliest maize in the south, clustering at 2000 BC, suggests that low-level maize agriculture was attractive to groups in a variety of settings, with alluvial settings apparently preferred. This strategy appears to have been relatively stable—with maize-using groups in the northern Southwest continuing to supplement diets focused on ricegrass, cheno-ams, and piñon for up to one thousand years. The reliability and predictability of maize was probably a major factor in its adoption, as the presence of this kind of stable resource would have been attractive to hunter-gatherers accustomed to exploiting wild grasses such as ricegrass and cheno-ams. Storability, ultimately allowing overwintering, would also have been a major impetus for adopting maize. I suspect that future research in settings amenable to early maize will continue to document these kinds of remains, as it is likely that maize spread via social networks from valley to valley and was adopted and integrated where possible.

Social networks were an important component of interaction during the Archaic period and would have facilitated both the introduction of the knowledge of maize cultivation (and later other cultigens) and the movement of actual maize and the technology associated with storing it. McBrinn (2005) used projectile point and textile data in an attempt to reconstruct Late Archaic social networks in the Mogollon Highlands and Jornada Mogollon region in the southern Southwest. She examined materials from a series of cave sites, including Tularosa, Cordova, and Fresnal rock shelters and found that all of these sites contained similar projectile points, indicating shared regional social networks. These networks would have provided information on the distribution and availability of wild resources, and would have facilitated the movement of ideas and innovations. This may account for the widespread presence of similar point styles throughout the region, as they would have been essential for the maintenance of these social networks.

A gradual increase in the number of sites with maize is seen over time. Rock-shelter sites with maize dates in the 1000 BC range suggest that maize storage became increasingly important as an overwintering strategy through time. In the meantime, open-air sites in alluvial settings begin to exhibit evidence of prolonged and more intensive occupation in the form of pithouses, storage pits, and middens. These are settings where maize productivity could have been enhanced using a number of agricultural production techniques, which would allow for more reliable and predictable diets that in turn led to settlement changes, most notably increased sedentism and storage, to enhance these adaptations. None of the sites dating to this 1000 BC time period were occupied year-round, however, and the initial and early use of maize appears to have been embedded in a seasonal foraging strategy across this entire region.

The suite of characteristics associated with Basketmaker II groups appears to have developed from these early maize-using adaptations, with groups later moving into regions such as the canyons and plateaus of northern Arizona and southeastern Utah once agricultural technology and maize strains were sufficiently developed to sustain an agricultural adaptation; certainly the reliance on maize indicated by the isotopic data and other studies support this (Matson and Chisholm 1991; Coltrain et al. 2007). The development of the Basketmaker pattern remains an important avenue for future research, but it is becoming increasingly clear that this pattern is not associated with the earliest use of maize agriculture in the Southwest.

In a 1994 article on the agricultural transition in the northern Southwest, Smiley remarked that "simple migration or adoption modes may be dangerously simplistic" (Smiley 1994:167). Twenty years later, that caveat holds true. New data have resulted in pushing back the agricultural transition to 2000 BC, nearly one thousand years earlier than dates associated with migration models proposed. Despite this fact, the questions being asked are still often framed in whether the data represent a migration of agricultural populations or diffusion of agricultural knowledge and techniques to the northern Southwest. It is likely that this is tied to the fact that we have not decoupled the initial use of maize from the later spread of maize. As described above, the initial use of maize appears to be substantially different from the later focus on maize, especially during Basketmaker times. The data presented in this chapter indicate that maize agriculture entered this portion of the Southwest in a variety of different ways, and the response to it varied depending on the circumstances surrounding the Archaic period occupants (or lack of them). It is thus time to start asking new questions of the data.

Note

1. Previous reports of early maize dating to 1660 BC from LA 107577 in dunes west of Bernalillo in the Albuquerque region are not included here because the remains were found to be grasses, not maize (Seymour et al. 1997).

The Southern Southwest 5

THIS CHAPTER EXAMINES THE TIME PERIODS immediately preceding, encompassing, and following the introduction of cultigens to the southern Southwest, which includes the Sonoran Desert of southern Arizona and the Chihuahuan Desert of southwestern New Mexico (figure 1.1, chapter 1). The time periods discussed in this chapter include the Middle (5000–1500 BC) and Late (1200 BC–AD 200) Archaic periods (following Huckell 1996a). As with the previous chapter, data from each of the regions where early agriculture has been found are presented on pre-agricultural Middle Archaic occupations, evidence for the earliest agriculture, and post-introduction occupations. Foraging adaptations continued in the Chihuahuan Desert region of southeastern Arizona and southern New Mexico long after maize agriculture was a major subsistence focus along the major river valleys of southern Arizona, so these differences in the integration of maize agriculture are discussed and evaluated. Data from two sites in northern Mexico, La Playa and Cerro Juanaqueña, are included in the discussion of post-introduction adaptations, along with their potential for documenting differences in identity among early farmer-foragers in the southern deserts.

History

The southern Southwest was referred to as the "southern sector" in Cynthia Irwin-Williams's Picosa Concept (1967), but is best known as the "Cochise Culture" first described by Sayles and Antevs (1941) based on their work at alluvial sites along Whitewater Draw in southeastern Arizona. They used stratigraphic data from eroding sites to define three separate occupations encompassing a time interval from the Early through Late Archaic periods (figure 5.1). The earliest, the Sulphur Springs phase (8000–5500 BC; Early Archaic period), was characterized by the first use of ground stone, slab milling stones and cobble manos, along with chipped stone tools such as scrapers and knives. No projectile points were found associated with these occupations by Sayles and Antevs, but further work at Early Archaic sites in the southern Southwest have documented the association of large stemmed points, similar to those found in the Great Basin and Mojave Desert, with Early Archaic period occupations (Huckell 1984; Roth and DeMaio 2015). The presence of ground stone and the lack of projectile points led Sayles and Antevs to posit that Sulphur Springs groups focused primarily on gathering versus hunting, which led to the idea that plant resources were extremely important to Archaic period foragers in the southern deserts from the earliest Archaic period times.

Sayles later defined an intervening phase between Sulphur Springs and the later Chiricahua phase that he called the Cazador phase (Sayles 1983). He separated it from Sulphur Springs based on the presence of projectile points in the artifact assemblage; however, further research has not supported the inference that this is a separate phase (Waters 1986), and both Sulphur Springs and Cazador are currently classified under the broader term Early Archaic (Huckell 1996a).

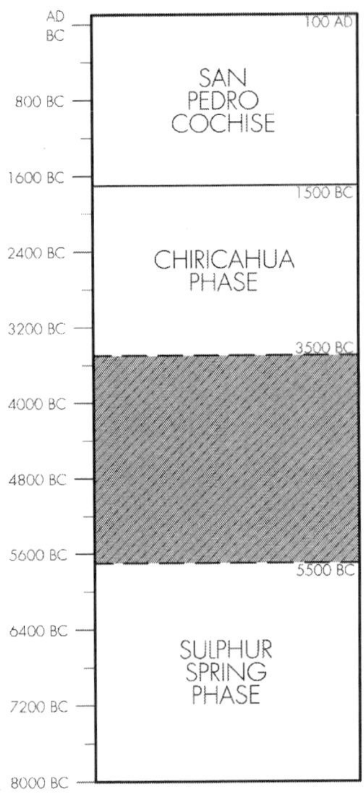

Figure 5.1. Cochise Cultural Sequence as proposed by Sayles and Antevs (1941). Figure by Ben Van Alstyne

The Chiricahua phase (3500–1500 BC) represents the Middle Archaic period and is perhaps most pertinent to agricultural beginnings, as it is during this time frame that agriculture is first found in the Sonoran and Chihuahuan deserts (see below). This phase has now been subsumed under the more general term "Middle Archaic period" (Huckell 1996a). Sayles and Antevs (1941; Sayles 1983) defined the Chiricahua-phase artifact assemblage as consisting of distinct projectile points with side and basal notching, bifaces, bone tools, and shallow basin metates. They identified several different site types, including base camps and specialized hunting and gathering camps. Further work at Middle Archaic sites has greatly expanded our understanding of this period.

The San Pedro Cochise period, now subsumed under the Late Archaic period, was first identified by Sayles and Antevs (1941), based on their work at both Whitewater Draw and in the Sulphur Springs Valley of southeastern Arizona. They dated this period to 1500 BC–AD 100, again based on geological data (Sayles 1983). San Pedro phase sites were identified primarily based on the presence of distinct San Pedro points, with side-to-corner notching, a triangular blade, and straight base. This point type essentially became the "type fossil" used to designate the Late Archaic period in the southern Southwest. The distribution of these points ranges from northern Mexico to north-central Arizona and from central New Mexico to central Arizona, but Sayles and Antevs (1941; Sayles 1983) confined their discussion of the San Pedro Cochise to sites in southeastern Arizona and southwestern New Mexico. Other artifacts associated with San Pedro sites included bifaces, scrapers, deep basin metates, manos, and the initial use of the mortar and pestle, which Sayles (1983) tied to changes in subsistence.

The first use of structures was also documented dating to the San Pedro phase (Sayles 1945). Excavations at Benson 5:10 near Fairbank, Arizona, yielded several shallow oval pits approximately two meters in diameter that Sayles considered to represent houses. Storage pits were also found at the site, which dated to 600–500 BC and were associated with what Sayles (1945) referred to as a "San Pedro Cochise" assemblage. A second site, Benson 8:3, yielded a similar shallow house floor with an internal pit. Because these sites were excavated before the advent of flotation analysis, no subsistence remains were found to allow for the interpretation that they were associated with early agriculture.

The notion of the San Pedro Cochise as representing a specific culture occupying the southern deserts has been pervasive in our understanding of Archaic period adaptations in this region, similar to the way that the Oshara concept has influenced interpretations in the northern Southwest. The presumption that the presence of San Pedro points is somehow related to the phase characteristics as documented by Sayles and Antevs (1941) and more recently by Mabry (2005a) has been important in the perception that the San Pedro phase represents dramatically different adaptations from those of the preceding Middle Archaic period. While recognizing that the Cochise represented substantial changes from the earlier Chiricahua phase (Middle Archaic period) occupations that they identified, Sayles and Antevs (1941) did not argue that these represented different groups, in large part because they documented stratigraphic continuity in these occupations. However others (e.g., Huckell 1995; Carpenter et al. 2002) have used the differences between San Pedro and Chiricahua phase occupations to support the inference that these represent two different groups of foragers and farmers, thus suggesting that San Pedro Cochise farmers migrated into the region.

Our knowledge of the Middle and Late Archaic periods in the southern deserts has expanded significantly in the last few decades, in large part because of major cultural resource management projects along the Santa Cruz River in the Tucson Basin and in the Jornada region of southeastern New Mexico. These developments have caused a substantive change in the way agricultural origins are viewed, especially along the major waterways of the Sonoran Desert. These projects have been done primarily along alluvial settings, however, so it is important to examine the variability present in the southern Southwest in order to determine the circumstances surrounding the introduction of agriculture to this region and the variability in the responses to the adoption of cultigens.

Environment

Agricultural beginnings in the southern Southwest occurred within a distinct environmental context. The region as defined here encompasses two deserts—the Sonoran and Chihuahuan. The Sonoran desert in the Basin and Range physiographic province of southern Arizona and northern Mexico is characterized by broad alluvial valleys separated by mountain ranges. Mountain erosion has resulted in the formation of pediments (bajadas) that have been covered with alluvial soils. Several rivers drain the alluvial basins, including the Santa Cruz, Salt, Gila, and San Pedro in southern Arizona. Springs occur in the adjacent mountains and were likely important prehistoric water sources, given the distribution of sites near them. There are two major rainy seasons in this desert, a winter season from November to March and a summer monsoon season from June to September (Dunbier 1968). The average frost-free growing season is 240 to 250 days. Vegetation is characterized by the Sonoran Desert scrub community and includes mesquite, paloverde, a variety of cacti, including saguaro and prickly pear, and grasses distributed across floodplain and bajada zones (Turner and Brown 1982). Agave and acorns are present in the mountains adjacent to the alluvial valleys, along with deer and, occasionally, bighorn sheep. Rabbits are available year-round, although their densities may fluctuate seasonally.

The Chihuahuan Desert of far southeastern Arizona and southern New Mexico is also part of the Basin and Range physiographic province, with environmental zones more widely spaced in the broad valleys (often referred to as "bolsons") than in the Sonoran Desert (figure 5.2). The Rio Grande is the only river found in this region, but springs and playas would have served as additional water sources (Doleman 2005). Most of the rainfall occurs during the summer monsoons from July through September and winter rainfall is highly variable (Miller and Kenmotsu 2004). Vegetation is dominated by semi-desert grasslands, and these desert grasslands were likely more extensive prehistorically, providing grama grass, yucca, and cacti, with screwbean and honey mesquite present along major washes. Mesquite-stabilized dunes are also present in some areas. Deer and pronghorn antelope are the primary large game present in this desert, along with rabbits and a variety of smaller species. Bison were present prehistorically, especially in the eastern portion of the region where grasslands would have supported them (Miller in press). Fish are available in the Rio Grande, but there is no evidence of Archaic period exploitation of them; however, this may be due to recovery techniques rather than absence of use. Miller and Kenmotsu (2004:208) argue that "climatic patterns [in the Chihuahuan Desert] created an uncertain, unpredictable, and geographically variable environment from year to year." Doleman (2005:131) also notes that resources are less predictable in the Chihuahuan Desert than in other portions of the Southwest and argues that this restricted the subsistence and settlement options of prehistoric hunter-gatherers in this area.

Figure 5.2. Chihuahuan Desert environment. Photo courtesy of Christine Ward

One fact that becomes apparent in evaluating data from the southern Southwest is that the vast majority of sites are open-air sites, and for the periods postdating the arrival of maize, they are primarily in alluvial settings. This contrasts with the northern Southwest, where evidence on early agriculture comes from rock shelters and more recently from open-air sites (see chapter 4). Geological factors play some role in this distinction, as there are fewer habitable rock shelters in the southern deserts, but both climate and differing land-use strategies likely also played a role in this distinction. It appears that alluvial settings were sought-after locales because of the presence of reliable water sources, abundant wild resources, and expanses of arable land. Upland occupations were primarily focused on canyons, with many of the sites located at canyon heads with access to a diverse range of resources and/or adjacent to springs. Often, but not always, these areas are away from rock outcrops with suitable shelters. It is also possible that rock shelters in the northern Southwest and Mogollon Highlands were preferred for overwintering because of the harsh winters; this was not a major concern in the southern deserts. The available data from sites like McEuen Cave and Fresnal Shelter, discussed below, indicate that rock shelters were sometimes used, and these occupations represent seasonal use with the integration of maize into a foraging economy similar to what has been documented in the Mogollon Highlands and northern Rio Grande.

The Middle Archaic Period

Middle Archaic sites have not been investigated as thoroughly as Late Archaic sites in the southern Southwest, and much of our knowledge is based on surface survey data supplemented with limited excavation. As a result, little subsistence data are available to fully evaluate the nature of foraging adaptations, and many inferences about subsistence practices have been based on artifact data, primarily the percentage of ground stone and hunting tools (projectile points). In addition, many Middle Archaic sites in alluvial settings have either been eroded as a result of river down-cutting (Freeman 1997a; Haynes 1968; Huckell 1996b; Waters 1986) or are deeply buried, sometimes under up to three meters of alluvium, making any kind of broad-scale exposure difficult (Gregory 1999; Whittlesey and Foster 2010). As a result, our knowledge of the Middle Archaic is sporadic, and caution must be taken in making regional interpretations from the existing database.

As with the northern Southwest, chronological issues are a problem for researchers studying the Middle Archaic period in this region, largely because of the broad time scales associated with the projectile point types used to classify sites as Middle Archaic. The major Middle Archaic projectile point types in the southern Southwest are Chiricahua, Pinto, and Cortaro points (figure 5.3). Chiricahua points have triangular blades, shallow side notches, and concave, eared bases. The original dating of the Chiricahua phase was accomplished via the geologic context of sites with Chiricahua points, which Sayles (1983) argued placed them in the 5000–3000 BC time range. However, Waters (1986) used radiocarbon dating to argue for a later chronology, from 3500 to 1500 BC. Chiricahua points were recovered from deposits with associated radiocarbon dates of 3500–2500 BC in the Picacho region north of Tucson (Bayham et al. 1986), supporting this later chronology.

As noted in chapter 4, Pinto points encompass considerable morphological variation and a broad temporal and geographic span, and their utility as a diagnostic marker both in the Great Basin and the Southwest has been challenged (Basgall and Hall 2000; Schroth 1994; Warren 1980). If the dates proposed by Basgall and Hall (2000) for Pinto points of 5/6000–2000 BC hold true for areas outside the Great Basin and Mojave Desert, then they establish a Middle Holocene occupation throughout southern Arizona. This idea is supported by the recovery of Middle Archaic components containing Pinto points and other Middle Archaic period projectile points associated with Early Archaic period stemmed points at some upland sites in the Tucson Basin and surrounding areas, suggesting continuity in occupation (Huckell 1984; Kaldahl 2001; Roth and DeMaio 2015). The only dated deposits with Pinto points in southern Arizona come from the Buried Dune site in the Picacho Basin (Bayham et al. 1986), which yielded dates of 3000–2500 BC.

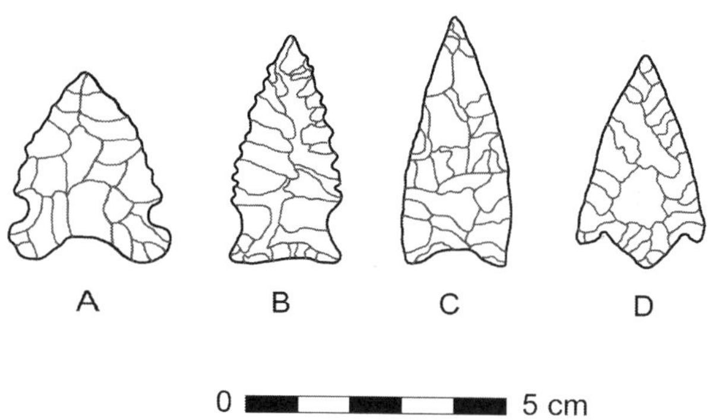

Figure 5.3. Middle Archaic period projectile points in the southern Southwest: A = Chiricahua, B = Pinto, C = Cortaro, D = Gypsum. Figure by Aaron Woods

Cortaro points, triangular concave-base points first defined by Roth and Huckell (1992), span the time period between the end of the Middle Archaic and the early Late Archaic periods (figure 5.3). Their distribution is confined to the southern Southwest. Dated deposits with Cortaro points from the Los Pozos and Las Capas sites along the Santa Cruz River in the Tucson Basin place their use from 2500–2000 BC (Gregory 1999; Whittlesey et al. 2010); however, they co-occur with Late Archaic projectile points at multiple sites throughout the southern Southwest, suggesting that they were used into the early portion of the Late Archaic period (Roth and Huckell 1992). Chronological issues remain a key problem in evaluating Middle Archaic occupations in this region, and additional research is needed to more fully establish the temporal range of these projectile point styles.

Despite these caveats, data on Middle Archaic period occupations are available that can aid in evaluating the adaptations represented by these groups and place the initial introduction of maize in context. These data have been greatly enhanced by contract archaeology projects, especially in the Jornada area.

Middle Archaic Period Occupations

Middle Archaic sites have been recorded in a wide variety of settings in the southern Southwest, including river floodplains and terraces, dunes, playa shores, mountain pediments, and mountains. Site types range from base camps, some of which were apparently reoccupied over long periods of time, to more specialized hunting and gathering sites. Some base camps contain dense artifact scatters with features including roasting pits and hearths, indicating a shift to some level of logistical mobility, with groups occupying sites in areas with abundant and reliable resources and often returning regularly to them. Roth (1992, 1995) has interpreted this patterned land use as indicating that Middle Archaic groups in the Tucson Basin moved within well-defined seasonal rounds. Structures have been found at some sites; these are generally small (two- to three-meter diameter) shallow, circular to oval features that likely represent brush structures (Bayham et al. 1986; Dart 1986; Halbirt and Copus 1993). The more specialized hunting, gathering, and lithic procurement sites found throughout the region likely represent activities of smaller task groups who returned to base camps. The overall pattern of land use indicates a generalized hunting and gathering adaptation, but data from different portions of the region show that groups were adapting to local resources. The nature of the environment and distribution of resources, especially water sources, conditioned land use and resulted in some of the differences in occupation observed throughout the region.

San Pedro Valley

Norman Whalen (1971, 1973) examined Middle and Late Archaic site distributions in the San Pedro Valley of southeastern Arizona, focusing on a one-hundred-square-mile section between the alluvial floodplain and the Whetstone Mountains (figure 5.4). Middle Archaic (Chiricahua) sites identified in his study included campsites with ground and chipped stone tools and hearths, and specialized hunting and gathering sites. All base camps were located near major stream channels or springs, a pattern consistent with the idea that hunter-gatherers focused on reliable water sources during the Middle Holocene. Whalen argued that both Middle and Late Archaic period groups were adapting to resources that crosscut the major environmental zones, especially mesquite and deer, with groups wintering at lower elevation riverine sites and then spending late summer and early fall in the adjacent foothills and mountains. He found that higher elevation sites were larger and argued that this was the result of seasonal aggregation that he interpreted as tied to the exploitation of piñon and deer. This seasonal pattern of land use is similar to what has been proposed for

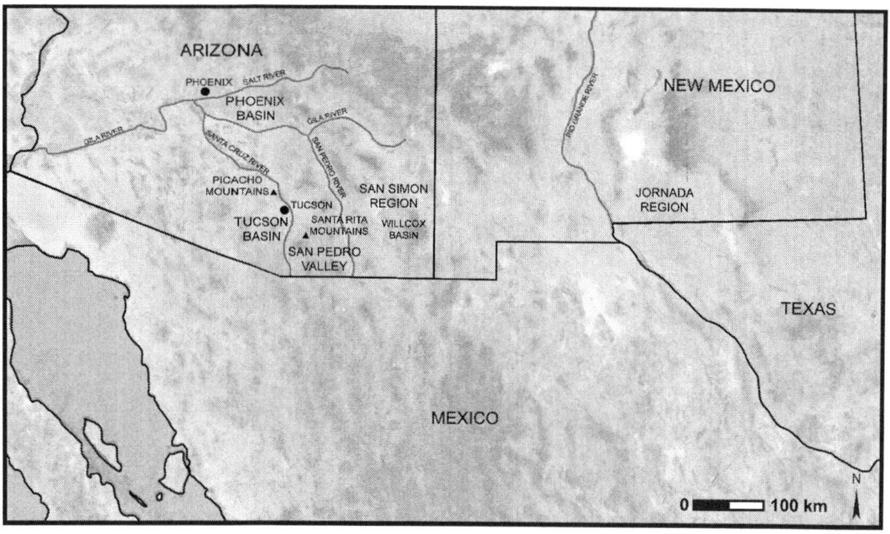

Figure 5.4. Regions of the southern Southwest discussed in this chapter. Figure by Ben Van Alstyne

other portions of the San Pedro River Valley (Phillips et al. 1993) and the Tucson Basin (see below).

Agenbroad (1966, 1970) reported on the Lone Hill site (AZ BB:10:17), a Middle Archaic period site located west of the San Pedro River Valley on the eastern pediment of the Santa Catalina Mountains. The site was a large base camp with numerous hearths, abundant ground stone, and a dense lithic scatter, including many Chiricahua and Pinto projectile points. Basin metates, some of which were heavily worn, were found in inverted positions on the site, and Agenbroad (1970) interpreted this as indicating that groups regularly returned to the site. The presence of a nearby cienega may have been the primary impetus for the intensive and repeated use of this site.

Tucson Basin and Surrounding Regions

Middle Archaic sites have been found widely distributed in all ecological zones in the Tucson Basin, with evidence of occupation of the upper bajada (mountain pediment), floodplain, and dunes (figure 5.4). Most of the data on the Middle Archaic period comes from upper bajada sites, most likely because of the surface visibility of these sites, as Middle Archaic floodplain sites are often deeply buried. Upper bajada sites are generally located near canyon heads or springs (Chavarria 1996; Dart 1986; Douglas and Craig 1986; Kaldahl 2001; Roth 1992, 1995, 1996), and data from excavated sites points to generalized hunting and gathering activities focused on large game, the abundant cacti present in this zone (including saguaro), and mesquite and other riparian resources present in large washes that cross-cut the bajadas. Dart (1986) found a small, shallow ephemeral structure at the La Paloma site (AZ BB:9:127) in the upper bajada of the Santa Catalina mountains and Kaldahl (2001) excavated several cache pits containing ground stone at the Rosetta site (AZ BB:9:248), again suggesting a shift to logistical mobility strategies. Middle Archaic sites have also been recorded on dune sites in the Avra Valley west of the Tucson Basin (Downum et al. 1986). Surface features consist of hearths or concentrations of fire-cracked rock associated with Chirichaua, Pinto, and Cortaro points; these sites may have been occupied because of the density and diversity of available plants on the dunes (Downum et al. 1986). The sites have both Middle and Late Archaic projectile points, suggesting continuity in site use over time.

Floodplain occupations are not as well documented as those in the upper bajada, primarily because of site visibility rather than a lack of occupation. Prehistorically, this zone contained mesquite bosques with adjacent grasslands that would have been attractive to hunter-gatherers. The most comprehensive data on the Middle Archaic period occupation of the floodplain come from the Los Pozos site (AZ AA:12:91), a multicomponent Middle and Late Archaic period site located along the Santa Cruz River above its confluence with the Rillito (Gregory 1999). Excavations of Middle Archaic period deposits located beneath two to four meters of alluvium revealed two stratigraphically distinct components. Only a small portion of the earlier component, buried beneath four meters of alluvium, was excavated; one pit and an associated occupation surface were found and charcoal from the pit dated to 3300–3000 BC. The upper component yielded several features including shallow pits, hearths, and oxidized surfaces in deposits dating to 2500–2000 BC. Cortaro and Pinto/San Jose points were recovered in separate areas of site, and Gregory (1999) argues that these two point types may represent temporally and/or functionally distinct occupations. Flotation samples from the features yielded cheno-ams, dropseed, tansy mustard, and mesquite, indicating a late spring and summer occupation focused on wild plant procurement. According to Gregory (1999), the data from Los Pozos point to short-term, episodic use of the floodplain by Middle Archaic period foragers. A maize cupule dating to 2500 BC was found during test excavations in a trench wall; this may represent very early maize use associated with the seasonal foraging adaptation (see discussion below).

A substantial Middle Archaic period occupation has been found north of the Tucson Basin in the Picacho Mountains area, where surveys documented Middle Archaic sites located in stabilized dune fields and on the bajada of the Picacho Mountains (figure 5.4; Czaplicki 1984; Bayham et al. 1986). Some large campsites were found in the dunes, with hearths, roasting pits, and dense scatters of chipped and ground stone. Excavations at two sites, the Arroyo site (AZ AA Z:3:28) and the Buried Dune site (AZ AA:3:16) provide further evidence of Middle Archaic use of this area. The Arroyo site had a midden deposit with a dense, diverse artifact scatter, clusters of fire-cracked rock, and a possible structure. The site yielded Chiricahua points, and radiocarbon dates place its occupation at 3600–2500 BC. The Buried Dune site contained fewer artifacts associated with several hearths and fire-cracked rock concentrations, and was interpreted as a short-term occupation dating to 3000–2500 BC. The site had San Jose/Pinto points and Shackley (1986) argued that these perhaps represent two different groups exploiting the dune fields during different seasons. Bayham and Morris (1986) tied the Middle Archaic period occupation to wetter conditions when more vegetation would have been present at certain times during the Middle Holocene, which would have attracted foragers from nearby areas and resulted in episodic use tied to climatic circumstances.

Another Middle Archaic occupation was investigated at the Tator Hills site (AZ AA:6:18) located on the bajada west of the Picacho Mountains (Halbirt and Copus 1993). Excavations revealed an occupation surface with roasting pits, hearths, and one small, oval structure that the excavators interpreted as a brush structure. Two Middle Archaic projectile points, including one Chiricahua point, were found on this surface, along with manos, metates, and a pestle. Pollen remains from the excavated features included mesquite, cholla, and cheno-ams, suggesting a spring and summer occupation.

Middle Archaic sites have also been identified in all ecological zones in the Phoenix Basin (figure 5.4). Sites have been identified primarily on surveys based on the presence of Middle Archaic projectile points and appear to represent seasonal movements tied to resource availability (Hall 2016). Some of these sites exhibit evidence of the repeated occupation of sites in resource-rich areas, especially in upper bajada and riverine settings. For example, work at the Last Ditch site (AZ U:5:33) on the bajada of the McDowell Mountains recovered a large number of thermal pits, two small oval structures, and an occupation surface dating to the Middle Archaic period, circa 2900–2000 BC (Hackbarth 1998, 2001; Phillips et al. 2001). Hackbarth (2001) interprets this as a seasonally reoccupied site that was used primarily for plant procurement and processing. Work at the Falcon Landing site (AZ T:7:419) in the western Phoenix Basin documented a similar pattern of intensive use of the bajada focused on wild plants (Hall and Wegener 2014; Wegener and Hall 2014). The Middle Archaic component is characterized by Chiricahua and Cortaro points. The site dates from the Middle Archaic through the Late Archaic period, circa 3300 BC–AD 400, suggesting continuity in site use over time, probably due to the presence of a nearby cienega and mesquite bosque (Hall and Wegener 2014; Wegener and Hall 2014).

Southeastern Arizona

Huckell (1984) reported on the Archaic period occupation of the Santa Rita mountains in southeastern Arizona south of the Tucson Basin (figure 5.4). The environment in this area included diverse plant communities including desert grasslands, chaparral, and oak-pine woodlands. The Middle Archaic occupation is represented by sites with Pinto/San Jose and Cortaro points that were seasonal occupations for gathering acorns and hunting deer.

Further east, Middle Archaic sites with Pinto and Chiricahua points have been found in the Willcox Basin (figure 5.4), but they represent short-term occupations of the bajadas and foothills because the Willcox Playa was empty for much of the Middle Holocene (Waters and Woosley 1990). Waters and Woosley (1990) reported on two episodes of lake refilling, one dating to 4000 BC and a second to 2000–1000 BC. This later episode is associated with a more substantial Late Archaic period occupation, discussed below. Gilman (1997) reported on Archaic sites found during surveys in the San Simon region of southeastern Arizona (figure 5.4). The Middle Archaic sites appear to represent relatively high levels of mobility, likely tied to the patchy environment and lack of reliable water sources in this area.

Southwestern New Mexico

Most of our knowledge of the occupation of southwestern New Mexico comes from extensive fieldwork done in the Jornada region (figure 5.4), especially work on Fort Bliss near El Paso, summarized in syntheses by Miller and Kenmotsu (2004) and Miller (in press). Small-scale contract surveys have also been done throughout the region, along with a more extensive survey associated with the border fence project in New Mexico's Boot Heel region (Turnbow et al. 2008). The use of this environment was conditioned by the availability of reliable water sources, which are sometimes widely spaced; the only perennial sources are the Rio Grande River and certain springs, but playas found across the landscape would have held water seasonally. As noted above, resource availability is unpredictable, and this factor structured Archaic-period adaptations throughout this desert environment (Doleman 2005).

MacNeish and Beckett (1987) and MacNeish (1993) placed the Jornada region into a separate "Chihuahuan Archaic Tradition" but recent syntheses (Miller and Kenmotsu 2004; Miller in press) establish that the similarities in technology and overall adaptations place it within the Southwestern Archaic. The phase designation proposed by MacNeish and Beckett (1987) is still widely used, with the Keystone phase (4000–1200 BC) representing the Middle Archaic period and the Fresnal and Hueco phases (1200 BC–AD 200) encompassing the Late Archaic.

Middle Archaic sites are identified primarily by the presence of projectile points, including San Jose/Pinto and contracting stem points (Gypsum,

Pelona). Gypsum points appear to be more common in this portion of the southern Southwest than in other areas, although they and similar styles like Elko Contracting Stem are found throughout the Southwest and into the Great Basin (figure 5.3). Like many of the other point styles discussed in this volume, Gypsum points are poorly dated in the Southwest. Many researchers classify them in the Middle-to-Late Archaic range, much like Cortaro points. Holmer (1986) reported that they dated from 2500 BC to AD 200 in the Great Basin, while Justice (2002:191) argues that they date from 2000 to 800 BC in the Southwest. Likewise, Huckell (1996a) argues that Gypsum points were no longer present in the Southwest after 1000 BC. Mabry (1998b:75) suggests, following Holmer (1986), that these points possibly "represent a specialized hafting technology developed for hunting artiodactyls."

Many of the Middle Archaic sites in the Jornada area are dispersed lithic scatters. The overall nature and distribution of sites reflects relatively high levels of residential mobility and low population densities (Miller and Kenmotsu 2004; Miller in press). Some sites in areas with higher resource densities and reliable water sources exhibit evidence of more intensive occupation and reoccupation during the Middle Archaic period. These often have features such as hearths, pits, and ephemeral structures that are typical of Archaic structures found elsewhere in the Southwest, as they are oval or circular and less than three meters in diameter. Wheaton (2013) has discussed these Jornada structures and argues that they represent brush huts.

Excavations at the Keystone Dam site near El Paso highlight the nature of these occupations. The site is located on a floodplain terrace with a nearby playa, which likely accounts for the evidence of more intensive and long-term use. Excavations revealed twenty-three ephemeral structures, of which three were excavated, and a variety of other features, including hearths and pits, dating from 2900–1500 BC (O'Laughlin 1980; Vierra 2009). O'Laughlin (1980) argues that this was a winter occupation with a focus on wild-resource procurement, including mesquite and cacti. Upham and MacNeish (1993) reported on the presence of tentatively identified maize pollen from Middle Archaic deposits at the Keystone Dam site. The context of these remains is not well established, however, and no macrobotanical remains were found, but it is possible that maize was also used at the site; if so, it represents the kind of low-level food production evident at other early sites in the region.

Data from excavated sites indicate that Middle Archaic groups were generally focusing on gathering mesquite, agave, yucca, and grasses and hunting deer and rabbits. The first burned rock middens, thought to be used primarily for baking agave and yucca, date to the Middle Archaic period (figure 5.5; Miller in press). These became important features on the landscape throughout the Middle and Late Archaic periods, indicating the continuing importance of wild-resource procurement even after the introduction of cultigens (Miller in press).

Playa lakes were present throughout the Jornada area and were apparently important locales for Middle Archaic period hunter-gatherers because of the presence of reliable water sources, increased plant density, and game. This is supported by the recovery of large numbers of Pinto and Gypsum points adjacent to the playas (Formby 1986). Playa lake stands are not well dated, but some of them may have been important locales for foragers during the Middle Holocene. Miller (in press) notes that site density throughout the Jornada region increased during the latter portion of the

Figure 5.5. Archaic period burned rock midden, Jornada region. Photo courtesy of Myles Miller

Middle Archaic period (Keystone phase). He ties this to favorable climatic conditions associated with a wet period starting at 2500 BC throughout the southern Southwest that resulted in the refilling of many playas, which would in turn have affected prehistoric foragers in this region because of the presence of reliable resources. The playa deltas would also have been ideal locales for planting maize using water-table techniques. The recovery of maize pollen in deposits dating to 2100 BC at LA 17280 along a playa in the Jornada region (discussed below) supports this (Graves et al. 2014).

Trends in Middle Archaic Period Adaptations

Data from these Middle Archaic occupations point to some adaptations that laid the foundation for changes that occurred during the Late Archaic period. Although Middle Archaic hunter-gatherers were responding to local conditions in different ways, some general trends are apparent. First, it appears that population increased. Although this is variable across the southern Southwest, it is relatively consistent in the recovery of both more and larger sites. Second, it appears that groups were practicing regular seasonal movements, resulting in patterned reoccupation of sites in productive settings with access to reliable water sources and abundant plant and/or animal resources. Wills (1992) has argued that this kind of patterned seasonal movement was a necessary precursor to the adoption of cultigens. Maize was introduced into these settings, making understanding the nature of these Middle Archaic period occupations critical for examining agricultural beginnings in this portion of the Southwest.

Agricultural Beginnings

The earliest evidence of maize in the southern Southwest dates to 2100 BC, which coincides with its appearance in the northern Southwest. Several researchers (Mabry 1998b; Whittlesey 2010) have suggested that the interval that encompasses the introduction of agriculture spanning the end of the Middle Archaic period (ca. 2000–1500 BC) to the beginning of the San Pedro phase (ca. 1200 BC) should be classified as a separate phase (referred to as the "unnamed interval" by Mabry and the "Sweetwater phase" by Whittlesey). However, given that we now have maize dating back to 2100 BC, indicating it was adopted during the Middle Archaic period, I have subsumed the interval from 2000 to 1200 BC under the Middle Archaic period.

Evidence for the initial use of maize in the southern Southwest comes from three deeply buried sites along the floodplain of the Santa Cruz

River in the Tucson Basin, one rock shelter site in the Gila Mountains in southern Arizona, an alluvial stream terrace in southern New Mexico, and possibly a playa shore in the Jornada region of southwestern New Mexico (table 5.1; figure 5.6). Given the depth of the deposits at the buried floodplain sites, it is highly likely that additional sites are present throughout the Tucson Basin and perhaps in surrounding alluvial settings, such as in the San Pedro Valley and along playa margins in southwestern New Mexico. Interestingly, no early maize sites have yet been found in the Phoenix Basin; despite extensive backhoe trenching along the floodplain during data recovery projects, the earliest agricultural components there date to the late Cienega phase (ca. 300 BC) (Hall in press). The reasons for the lack of evidence of early agriculture in the Phoenix Basin are not entirely clear, as the Salt and Gila Rivers would have provided ample locations for both floodwater and irrigation farming, and the resource base is similar to that found in the Tucson Basin. The ecological zones in the Phoenix Basin are more widely spaced than those in the Tucson Basin, making logistical mobility more of a challenge; it is perhaps for this reason that the earliest farming sites are not located in this area. However, it is also possible that this is simply a factor of site visibility and that similar early agriculture components will eventually be uncovered at deeply buried sites in the Phoenix area. Graves et al. (2011) reported on the recovery of several Early Archaic period structures at the Cashion site (AZ T:11:94) on the lower Salt River

Table 5.1. Sites with Earliest Maize Dates in the Southern Southwest

Site	*Dates (Radiocarbon Years BP)*	*Calibrated BC Dates/ Intercept*	*Reference*
LA 170280	3660 ± 30	2135–1945 BC/2030	Graves et al. 2014
Los Pozos	4050 ± 50	2860–2470 BC/2590	Gregory 1999
	3340 ± 60	1770–1460 BC/1630	Gregory et al. 2007
	3300 ± 80	1770–1420 BC/1590	
	3230 ± 50	1620–1420 BC/1500	
Clearwater	3690 ± 40	2200–1960 BC/2080	Thiel and Mabry 2006a
	3650 ± 40	2140–1910 BC/2020	
Las Capas	4930 ± 30	3700–3650 BC/3660	Vint 2015
	3990 ± 30	2570–2460 BC/2520	Vint 2015
	3670 ± 40	2200–1940 BC/2050	Whittlesey and Foster 2010
	3440 ± 40	1880–1650 BC/1775	Vint 2015
McEuen Cave	3690 ± 50	2270–1940 BC/2080	
Montoya Site	3640 ± 30	2080–1870 BC/2020	Laumbach n.d.
	3610 ± 30	2060–1880 BC/1975	

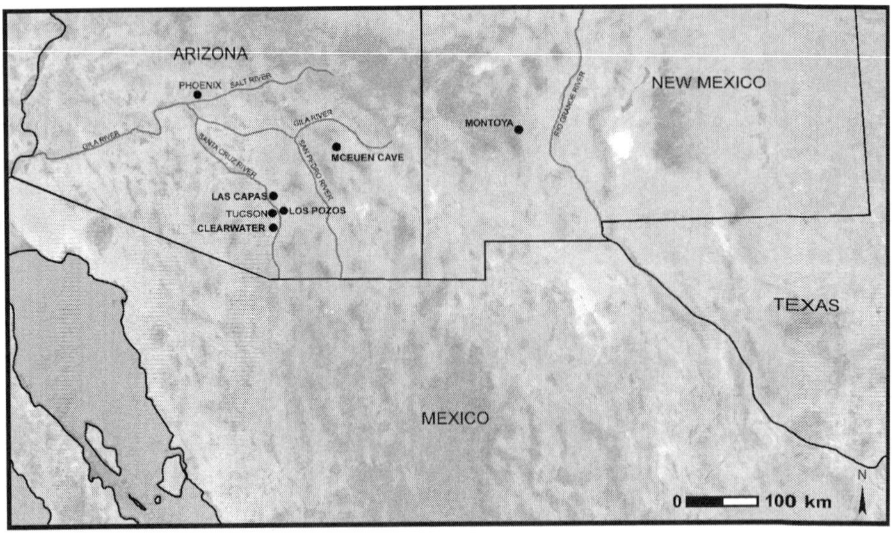

Figure 5.6. Sites with earliest maize in the southern Southwest. Figure by Ben Van Alstyne

floodplain and suggest that other Archaic period occupations may be present in this area. There is also evidence for Early Ceramic period (Red Mountain phase) occupations in the Phoenix Basin, so the absence of early agriculture in this area should be a topic of continuing investigation.

Tucson Basin Sites

Excavations at the Clearwater site (AZ BB:13:6) on the west side of the Santa Cruz River in the southern portion of the Tucson Basin revealed a cluster of ephemeral pithouses and numerous pits, including one roasting pit and a bell-shaped storage pit, with maize from the features dating to 2100 BC (table 5.1; figure 5.6; Mabry 2006; Thiel and Mabry 2006b). These remains were found in alluvial deposits 1.5 meters below ground surface. Projectile points associated with this occupation included San Jose/Pinto and Cortaro points.

Perhaps as surprising as the early maize remains, seven untempered, low-fired ceramic sherds representing six vessels were recovered from the features at the Clearwater site. These represent the earliest ceramics in the Southwest (Thiel and Mabry 2006b), and are what Heidke (1999, 2005) refers to as "incipient plainwares" representing the *in situ* development of pottery in the Tucson Basin. Incipient plainwares had been found at later San Pedro– and Cienega-phase sites, but none had been found in Middle Archaic deposits before. Heidke (2006) suggests that the sherds are from small bowls or cups and indicate specialized rather than domestic use. Ground stone and pigment from these deposits point to pigment processing, a practice that became more common during the succeeding San Pedro and Cienega phases (Adams 2015).

One irrigation canal was found at Clearwater dating to 1500 BC, and Thiel and Mabry (2006b:22.4) note that "this is the oldest known canal north of central Mexico" and appears to represent the independent development of irrigation technology. Its date establishes that irrigation was quickly added to the suite of farming techniques used in this portion of the southern Southwest.

Maize dating to 2100 BC has also been recovered from deeply buried Middle Archaic period features at the Las Capas site (AZ AA:12:111) located on the floodplain near the confluence of the Santa Cruz River and Cañada del Oro Wash in the Tucson Basin (figure 5.6). A small portion of a stratum dating to the Middle Archaic period was excavated during the expansion of a sewage treatment plant, but because of its depth of nearly four meters below ground surface, only a small area was exposed, resulting in the recovery of numerous pits, including several hearths, roasting pits, and a bell-shaped storage pit with associated occupation surfaces (Whittlesey and Foster 2010). Cortaro points were the only identifiable projectile points associated with this occupation. Maize from one roasting pit yielded a date of 2100 BC (table 5.1), and mesquite, cheno-ams, and saguaro were also recovered from a flotation sample from this pit (Whittlesey and Foster 2010). These data suggest that the occupation is associated with seasonal foraging activities.

Excavations of another portion of the Las Capas site conducted to document the extent and nature of the irrigation canal system recovered several early maize specimens with the oldest dating to 3700–3600 BC (table 5.1; Vint 2015). The oldest maize dates were from disturbed contexts, however, and the nature of the occupation associated with them is not clear. Other early maize was recovered from deeply buried deposits with several pits associated with Cortaro points, with dates ranging from 1800–1400 BC (table 5.1).

A maize cupule dating to 2500 BC was recovered at the Los Pozos site during test excavations (figure 5.6). This cupule was found in a trench wall associated with the Middle Archaic deposits. Gregory (1999:118) remarked that there were several problems associated with this sample, including the lack of other maize remains in any of the other flotation samples, the fact that the sample was too small to split so the results could not be reproduced, and the fact that the 13C/12C ratio had to be estimated. He referred to the date as "enigmatic" (Gregory 1999:118), but given the recovery of maize at several other floodplain sites dating to 2100 BC, it is possible that this represents maize use on the floodplain dating back to 2500 BC.

Excavations in another small portion of the Los Pozos site referred to as the Sweetwater Locus uncovered two shallow, oval structures associated with numerous pits, including a bell-shaped storage pit, and a low artifact density (Gregory et al. 2007). Maize cupules from the features dated to 1700–1400 BC (table 5.1) and other macrobotanical remains included saguaro, cheno-ams, and mesquite, again suggesting that the occupation was associated with summer foraging activities. No diagnostic projectile points were found in these deposits. Gregory et al. (2007:83) suggests that this locus represents "multiple discontinuous occupations." Los Pozos does not have an extensive San Pedro phase component, which immediately follows the initial use of maize in this area; this may be due to changes in the floodplain that made this portion of the floodplain less attractive, so groups moved to more stable sections with abundant resources (Gregory and Nials 2005). A substantial Cienega phase occupation is present, however, suggesting that groups returned to the site when the floodplain stabilized.

Gila Mountains

Maize dating to 2100 BC was recovered during excavations at McEuen Cave (AZ W:13:6), located in the Gila Mountains outside Safford, Arizona (figure 5.6). This site is in an oak-juniper woodland at an elevation of 4,400 feet in a transitional zone between the Sonoran Desert and Mogollon Rim. The cave was originally excavated by Bryan Cummings, and perishables from this early work have been dated to the Cienega phase (Moreno 2000). Subsequent excavations were conducted at the site in the 1990s (Huckell et al. 1999). These excavations revealed stratified deposits dating from the Middle Archaic through the Cienega phase of the Late Archaic period, with limited evidence of use during later periods. The results of this project have not been published, but some information is available on the Archaic period occupation. Projectile points from the Middle Archaic deposits included Chiricahua and Cortaro points. These deposits contained maize, and one kernel dated to 2100 BC (table 5.1). Other plant remains associated with the maize included oak, mesquite, yucca, cheno-ams, prickly pear, hackberry, and grasses. No formal storage facilities (e.g., pits or cists) were found, and it appears from the recovered assemblage that the site represents repeated seasonal use by Middle and Late Archaic groups, who integrated maize into their foraging economy.

There is evidence for continuity in occupation from the Middle to the Late Archaic, and no dramatic changes in the artifact assemblage were noted. This suggests continuity in land-use strategies over time.

Southwestern New Mexico

Two maize cobs dating to 2000 BC were recovered at the Montoya Site, a Mogollon pueblo located in the Rio Alamosa drainage of southern New Mexico (figure 5.6; table 5.1; Laumbach n.d.). These cobs were found at the base of a one-by-two-meter excavation unit located on a stream terrace. The majority of this Archaic period component was not excavated, so no other evidence is available on this find to contextualize early maize use. Like the other sites with limited evidence of maize, its significance lies primarily in documenting the presence of maize by 2000 BC at sites in a broad range of alluvial settings, indicating that by this time Archaic period hunter-gatherers were regularly experimenting with maize.

Work on Fort Bliss in the Tularosa Basin in southwestern New Mexico has uncovered a large open-air site along a playa with maize pollen that may also represent early maize use in the southern Southwest. LA 170280 is a Keystone phase (Middle Archaic) site with several clusters of hearths and pits along with abundant ground stone (Graves et al. 2014). The excavators argue that brush huts were probably present on the site. Maize pollen was recovered from one hearth with charcoal dating to 2100 BC (table 5.1). Two contracting stem (Gypsum/Augustin) points were found at the site. Other similar Late Archaic sites were recorded along the playa, and these may represent seasonal camps where maize was grown. As noted above, these are areas where early maize would be expected, given the resource density around playas, the presence of reliable water sources, and the potential for water-table farming.

Summary of Agricultural Beginnings

Data to examine agricultural origins in the southern Southwest come primarily from alluvial settings across the region. Unfortunately, most of these sites are deeply buried, so only small components have been excavated. The available data suggest that Middle Archaic foragers were seasonally exploiting the floodplain and perhaps playa edges to gather summer annuals and other plants such as mesquite, with the repeated use of some sites in productive settings. Maize was apparently adopted into this seasonal foraging strategy (Gregory 1999; Roth and Freeman 2008). Alluvial settings where floodwater and water-table farming could be practiced appear to have been preferred for integrating maize agriculture into the subsistence regime. As with the settings for the earliest maize in the northern Southwest, in general these are locales that fit with expectations of where hunter-gatherers would initially incorporate a new subsistence item.

Clearwater and Los Pozos both have small, ephemeral structures and processing features, and all three sites with early maize in the Tucson Basin have bell-shaped storage pits, supporting evidence discussed in chapter 4 that this storage technology was adopted along with maize. The projectile point types associated with these early sites include Pinto/San Jose, Chiricahua and Cortaro points, and Gypsum points in the Jornada region. Cortaro points have consistently been found at all of the early sites that have yielded diagnostic points except LA 170128. Cortaro points are found at Middle Archaic sites throughout the southern Southwest, often with later Late Archaic components, and their presence at these sites with the earliest maize further suggests that maize was adopted by local hunter-gatherers in the region.

The Late Archaic Period

The Late Archaic period (1200 BC–AD 200) witnessed substantial changes from the previous Middle Archaic period in some portions of the southern Southwest in response to the integration of and subsequent increased dependence on maize agriculture. This section focuses primarily on the San Pedro phase (1200 BC–800 BC), which represents the initial period when changes associated with maize agriculture start to appear and gel. The succeeding Cienega phase (800 BC–AD 200) is discussed in terms of the continuing impact of agricultural production on farming groups in the southern Southwest. However, the Cienega phase, like Basketmaker II in the northern Southwest, represents a time period that postdates the initial use of maize by a considerable amount of time, so the characteristics of this period are used primarily to document trends in adaptations rather than for its potential to help us understand the transition from foraging to farming in this region.

As noted in chapter 2, Huckell (1995) used the profound changes in occupation observed in the San Pedro and Cienega phases in the San Pedro River Valley and Tucson Basin to argue that these phases should be classified as "Early Agricultural" rather than as Late Archaic. This term has been variously accepted by Archaic period researchers; not surprisingly, it is used predominantly by researchers in southern Arizona, where it is perhaps the most parsimonious with the data. Because the use and integration of cultigens varied so much across this region, I have opted to continue to use the term "Late Archaic," although recognizing the extensive variability in the impact of agriculture among these different groups.

San Pedro points are the primary diagnostic projectile points used to classify sites as dating to the San Pedro phase, although many of the excavated sites have also yielded radiocarbon dates. These points, which date from 1200 BC–AD 50, remain a hallmark of the San Pedro phase throughout the southern Southwest and into portions of the northern Southwest such as the Mogollon Highlands (figure 5.7). Empire points are serrated leaf-shaped to triangular-blade stemmed points with rounded bases that date to the early San Pedro phase (1200–900 BC) (figure 5.7; Sliva 2015; Stevens and Sliva 2002). These points have been recovered from the Cienega Valley in southern Arizona and in excavated deposits at Las Capas. Sliva (2015) has recently proposed changes to the San Pedro and Empire point typologies, identifying a large number of subtypes for each main type based on notch and base characteristics. Given the fact that other researchers have not yet adopted these subtypes and instead follow the more inclusive categories presented by Justice (2002), the points discussed here are classified under the more general categories of San Pedro or Empire points.

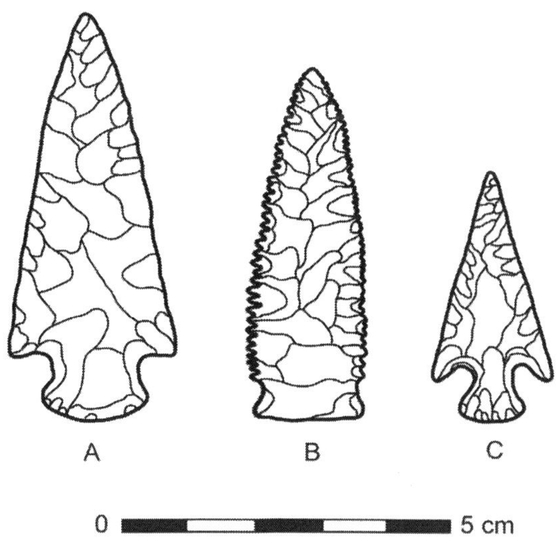

Figure 5.7. Late Archaic projectile points in the southern Southwest: A = San Pedro, B = Empire, C = Cienega. Figure by Aaron Woods

Changes observed during the San Pedro phase in some portions of the southern Southwest are substantial and include the presence of structures, large bell-shaped storage pits, a variety of other pits, and abundant maize, indicating that maize farming was an established part of the subsistence base by 1200 BC (Mabry 2005a). The structures generally are oval to circular, about three meters in diameter, and represent bent-pole construction (figure 5.8). Some sites, such as Las Capas in the Tucson Basin, contain large numbers of structures and features, indicating relatively intensive occupation. Changes in the artifact assemblage are also apparent, with the use of ceramic figurines, cornucopia-shaped ceramics, and shell jewelry.

A variety of agricultural strategies were used by these Late Archaic groups, including floodwater farming, water-table farming, and possibly alluvial fan farming using runoff from the bajadas (Mabry and Doolittle 2008). Irrigation agriculture has been well documented in the Tucson Basin; other than the canals found along the Zuni River on the Colorado Plateau (see chapter 4), these represent the earliest irrigation canals in the Southwest and are thus far confined to the Tucson Basin in the southern deserts (Mabry 2002, 2006, 2008).

These changes do not occur in all portions of the southern Southwest, however; other regions, especially the Chihuahuan Desert, exhibit long-term continuity in foraging adaptations, sometimes with the addition of maize agriculture, as has been documented for other Late Archaic sites in the Mogollon Highlands and northern Rio Grande region (chapter 4). These variations in adaptations indicate that maize was differentially adopted, and the impact of the adoption varied, with some areas exhibit-

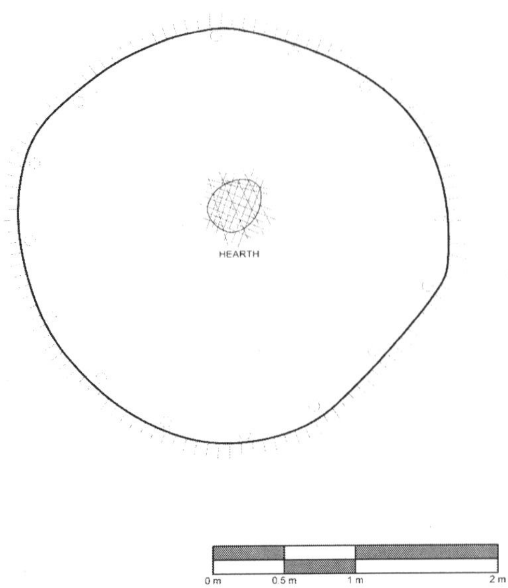

Figure 5.8. Example of an early structure in the southern Southwest. Figure by Ben Van Alstyne

ing substantial maize use associated with increased sedentism and the use of new techniques such as irrigation, and others remaining foragers and forager-farmers. This variation is described below and sets the stage for examining variability in the impact of agriculture in this region.

Tucson Basin

Late Archaic sites have been found in all major ecological zones in the Tucson Basin, but our understanding of this period comes primarily from alluvial settings along the Santa Cruz River and its tributaries, where the majority of excavated sites are located. Four San Pedro phase sites have been excavated on the floodplain of the Santa Cruz River and its tributaries, and two other sites, Wetlands AZ AA:12:90 (Freeman 1997a) and Solar Wells AZ AA:12:105 (Mabry 1990), have yielded San Pedro phase features (figure 5.9).

The first site to be documented was the Milagro site (AZ BB:10:46), located on a terrace bordering Tanque Verde Creek in the eastern Tucson Basin (figure 5.9). Excavations done there by the Huckells in the 1980s and early 1990s recovered pithouses, bell-shaped storage pits, hearths, and numerous other pits, with radiocarbon dates on maize placing the occupation at 1200–1000 BC (Huckell and Huckell 1984; Huckell 1990; Huckell et al. 1995).

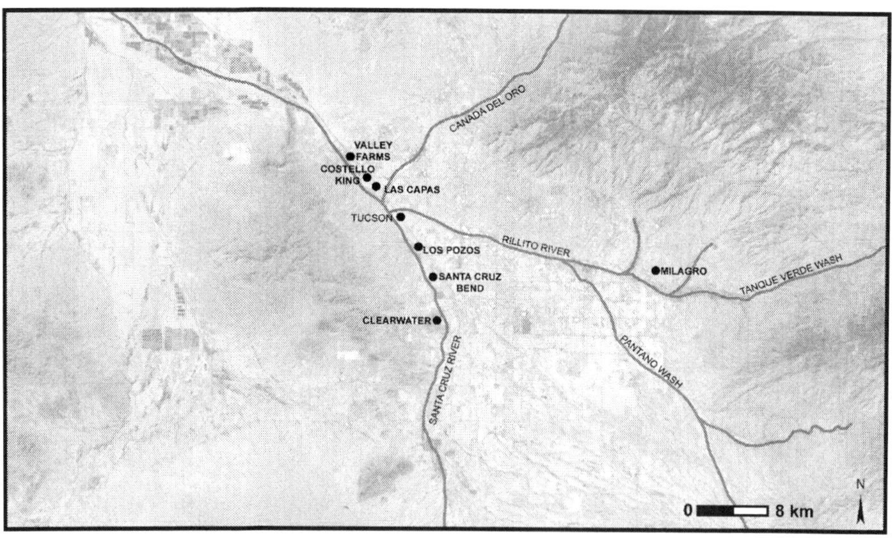

Figure 5.9. Late Archaic period sites in the southern Southwest. Figure by Ben Van Alstyne

Artifact density was low at Milagro; artifacts included San Pedro projectile points, numerous bifaces, a small amount of ground stone, clay figurine fragments, and shell. Maize was found in all flotation samples from the features, along with a range of wild resources (e.g., saguaro, mesquite, grasses, juniper, agave, and a possible acorn shell), including eight hundred charred amaranth seeds recovered from one pit. The site is ideally located to take advantage of several vegetative communities, including riparian, paloverde-mixed cacti, desert grasslands, and adjacent mountains, which explains the diversity of wild resources (Huckell et al. 1995).

Huckell has remarked on the large storage capacity of some of the bell-shaped pits at Milagro and has used that along with the ubiquity of maize recovered from the flotation samples to argue for high maize dependence by these early San Pedro phase groups. Huckell et al. (2002) did a computer simulation of storage capacity and argued that up to six thousand ears of corn could be stored in one of the larger pits. Some large bell-shaped storage pits were found clustered away from the structures, and Huckell and Huckell (1984) suggest that this might represent communal storage. Although Huckell et al. (1995) postulate a year-round occupation at Milagro, based on the labor invested in house construction, the large capacity of the storage pits, and the presence of spring and summer annuals, the ephemeral nature of the structures and the low artifact density indicate that it most likely represents a seasonal occupation (Roth 2016).

The Costello-King site (AZ AA:12:503) is located on the western edge of the Santa Cruz River floodplain to the north of the Las Capas site (figure 5.9; Ezzo and Deaver 1998). Backhoe trenching prior to construction of a waste management facility revealed 181 features, including one small oval pit structure, hearths, bell-shaped storage pits, basin pits, and a dog burial in deposits that dated to 1200–800 BC (Ezzo and Deaver 1998; Ezzo and Stiner 2000). Two irrigation canals were discovered associated with several hearths and pits with maize dates of 1000–900 BC. The only recovered plant remains were maize and several cactus seeds, along with abundant cheno-am pollen, and the tool assemblage indicates a focus on plant processing. Ezzo and Deaver (1998) argue that this was a seasonally occupied agricultural camp.

The Valley Farms site (AZ AA:12:736) is located on the floodplain one-half mile east of the Santa Cruz River (figure 5.9). Test excavations and data recovery were done in five study areas at the site covering an area approximately 1.3 miles long, and deeply buried Archaic period deposits were found in four of these (Wellman 2008). The Archaic period deposits extended to a depth of three meters below ground surface and date from 1200–800 BC, with the majority dating from 1200–1000 BC.

Excavations revealed 308 Archaic period features spread throughout the study areas (Thurtle et al. 2008). One oval, three meter-diameter pit structure and a second possible pit structure were found in Study Area 1 and these were the only Archaic period structures identified at the site. Other features included bell-shaped storage pits, roasting pits, hearths, and a range of other pits, along with oxidized surfaces, some of which are associated with postholes that may represent ramadas, and several burials. Flotation samples yielded abundant maize along with a wide range of wild resources, including saguaro, prickly pear, cholla, mesquite, and cheno-ams. The recovery of cattail pollen indicates that a cienega was present nearby.

Several spatially distinct clusters of features were present across the site, usually consisting of several roasting pits and storage pits, along with hearths and basin pits. Roth and Wellman (2001) interpret these as representing family groups who moved seasonally to the floodplain to farm, gather summer annuals, mesquite, and cienega resources, and hunt. The caching of material in storage pits suggests that some seasonal movement away from the site occurred.

The primary occupation of Las Capas (AZ AA:12:111), which has yielded early maize (discussed above), dates to the San Pedro phase (figure 5.9). Excavations have been done at three separate loci at the site; one along the Interstate-10 corridor dating primarily to the latter portion of the San Pedro phase (Mabry 2008), a second locus excavated prior to the expansion of a sewage treatment plant dating to the early portion of the San Pedro phase (Whittlesey et al. 2010), and a third excavated locus encompassing an irrigation canal system and associated agricultural fields along with features dating from the Middle Archaic period through the late San Pedro phase (Vint 2015; Vint and Nials 2015).

Excavations at the sewage treatment plant locus concentrated on a stratum dating to the early San Pedro phase, circa 1200–1000 BC (Whittlesey et al. 2010). This stratum yielded pit structures, extramural work surfaces, hundreds of pits (bell-shaped storage pits, thermal pits, basin, and cylindrical pits), irrigation canals, and burials (Hesse et al. 2010). Some of the larger structures were built using perimeter posts, indicating more labor was invested in house construction than in the more typical oval bent-pole structures, a trait generally associated with increased sedentism (Diehl 1992). San Pedro and Empire points were found associated with this occupation.

Work along Interstate 10 concentrated in a stratum dating to the late San Pedro phase, 900–730 BC (Mabry 2008; Vint 2015). Features found in this area included pit structures, specialized storage structures, bell-shaped storage pits, human and dog inhumations, occupation surfaces, and an array of processing and other extramural pits. Canals were found in all strata and could be traced across the site; the canal system and associated fields have been further defined during recent investigations at the site (Vint and Nials 2015). Plant remains from the late San Pedro phase component showed the same mixed farming-foraging adaptation as seen in the early San Pedro phase occupation at Las Capas, with abundant maize, mesquite, cheno-ams, and cacti (Diehl 2005b, 2015). San Pedro points dominated the projectile point assemblage and Sliva (2005) notes that in general the chipped stone assemblage remained stable through time, with the addition of several new artifact types including stone hoes and cruciforms. Ground stone from the site showed a continued focus on pigment processing throughout the San Pedro phase (Adams 2015).

Sliva (2015) argues that the presence of Empire points at Las Capas in deposits dating to the early San Pedro phase from 1200–1000 BC represent the presence of a distinct group of farmer-foragers who were later replaced by other groups bearing San Pedro points. She and Heidke (2015) use these projectile point differences and distinct characteristics in clay figurines from the early and late San Pedro components at Las Capas to support the idea that males migrated into the Tucson Basin, bringing maize and San Pedro points with them (see chapter 2). The direct equation of projectile points and ethnicity has been challenged and dismissed by many researchers (e.g., Sackett 1977), and Heidke (2015) notes that the changes in figurines could be associated with ritual changes tied to an increased emphasis on maize, so they are not necessarily ethnic markers. The fact that Empire points have been found in very restricted contexts, with the only known examples in the Tucson Basin from Las Capas, may be tied to individual innovation rather than group-level identity distinctions. Given what is known of hunter-gatherer behavior (see chapter 3), the presence of non-local males at Las Capas is much more likely to represent intermarriage between foraging groups than a migration of ethnically distinct males into the region. If the intermarrying males were participating in extended social networks, this may account for some of the variability in Empire and San Pedro points demonstrated by Sliva (2015).

Overall feature density was much higher at Las Capas than at any of the other San Pedro phase sites excavated in the Tucson Basin thus far and this portion of the floodplain appears to have been particularly productive for farming and foraging because of the high water table and extensive amounts of arable land (Gregory and Nials 2005; Nials 2008). However, its proximity to major tributary drainages also resulted in it experiencing several major flooding events that influenced the location of the settlement and resulted in movement away from Las Capas during the Cienega phase (Nials 2008).

Mabry (2008) has argued that Las Capas is unique in terms of its duration and continuity of occupation. He ties this to its location, arguing that groups were tethered to this area because of the favorable floodplain environment and irrigation practices. Although he sees Las Capas as being almost continuously occupied for a span of four thousand years, he argues that the locus of habitation shifted over time in response to floodplain changes, resulting in settlement drift (Mabry 2008). Whittlesey (2010) argues that the remains represent "episodic and recurrent use" with brief episodes of occupation punctuated by movement due to recurrent flooding. She uses the term "recurrent sedentism" to denote that groups maintained a focus on the "place" of Las Capas despite repeated movement, perhaps because of its lush setting and access to arable land. It is apparent from these data that Las Capas represents a different kind of occupation from that represented at the other San Pedro phase sites investigated in the Tucson Basin thus far, especially in terms of size and the number of features. It is not immediately apparent why it is so different, although the prime location for farming and wild-resource availability was clearly one reason for the intensive occupation. It is also surprising that there is no evidence of agricultural intensification evident in the macrobotanical, faunal, or artifact assemblage, given the expansive irrigation network and the intensity of site occupation. It represents one part of the spectrum of Late Archaic adaptations present in this region.

In addition to the excavated San Pedro phase sites on the floodplain, some information on the Late Archaic occupation of the upper bajada is available, primarily as a result of surveys and limited excavations (Dart 1986; Douglas and Craig 1986; Roth 1992, 1995, 1996). As with the Middle Archaic occupations, upper bajada sites consist of clusters of artifacts and features that are distributed primarily within the paloverde-mixed cacti vegetation zone, usually adjacent to water sources. These base camps often show evidence of use from the Middle through the Late Archaic periods, and the use of the upper bajada does not appear to have changed substantially over time, representing seasonal use for hunting and gathering activities. Evidence for more specialized, task-oriented use is present during the Cienega phase, perhaps associated with the increased sedentism discussed below.

Cienega phase (800 BC–AD 200) occupations are well documented in the Tucson Basin, including several large Cienega phase sites (Los Pozos, Santa Cruz Bend, and Clearwater) with hundreds of houses and pits (figure 5.9; Gregory 2001; Mabry 1998a; Thiel and Mabry 2006a). Cienega phase floodplain sites are characterized by increasingly formal structures and storage pits, burials, irrigation canals, and further elaboration of the artifact assemblage, including the addition of ground stone artifacts such as stone trays and the use of Cienega points, a corner-notched point with a triangular blade and expanding stem that Justice (2002) argues developed from the San Pedro point (figure 5.7). Gregory and Huckell (1998) found that structures remain oval to round in shape during the Cienega phase but increased in floor area and exhibited more formal roof and wall construction. This phase also witnessed an increase in the use of intramural over extramural storage features (Wöcherl 2005). These data indicate that groups were becoming more sedentary, although the intensity and duration of occupation remains a topic of debate (Diehl and Gregory 2002; Diehl and Waters 2006; Gregory 2001; Mabry 1998a). Three sites in the Tucson Basin (Clearwater, Santa Cruz Bend, and Wetlands) have large pit structures that have been interpreted as communal structures; these again support the inference that Cienega phase occupations were becoming increasingly sedentary and perhaps exhibit increased social complexity.

Maize, squash, and cotton were cultivated during the Cienega phase, and wild tobacco was also used. Surprisingly, subsistence data indicate that groups continued to practice a mixed farming-foraging adaptation with limited evidence of intensive agricultural production, with the exception of the canals (Diehl 2005b; Diehl and Waters 2006). The large storage capacity of the storage pits has been used to argue for extensive agricultural production, but this is not supported by the paleobotanical data. Diehl and Waters (2006) have argued that the fact that pit storage was a highly risky technique, with up to 30 percent loss, limited the ability to intensify beyond a certain extent. Many of the storage pits contain cached materials, especially ground stone, so they were not all used for food storage. Dean (2005) compared faunal assemblages from San Pedro, Cienega, and later Pioneer period Hohokam sites and found that the faunal remains from the Late Archaic sites did not represent sedentary components and that site use intensity did not increase until the Hohokam period. She sees this as indicating that early farming sites were seasonally occupied with low population densities.

The seemingly conflicting data—ubiquitous maize, large storage capacity, canal irrigation—versus the paleobotanical and faunal evidence of farming-foraging dietary stability may call for a more nuanced examination of agricultural intensification and its archaeological correlates. Unfortunately, isotopic data are not available from the burials to examine maize dependence; however, the increase in the number of burials recovered from Cienega phase contexts, the increasing formality of structures, the large storage capacity and use of specialized storage structures, and the construction of communal structures at some sites all point to some significant cultural changes, likely resulting from the interplay of social demands as groups became more sedentary and focused more on canal irrigation and the risks and demands of living on a dynamic floodplain in a desert environment.

Phoenix Basin

As noted earlier, early maize has not been recovered along the Salt and Gila Rivers in the Phoenix Basin, although this could be a result of site visibility factors. No deeply buried San Pedro phase sites with maize have been found either, although San Pedro phase components have been identified at numerous hunting and gathering sites in the Phoenix Basin (Hall in press). For example, both the Last Ditch and Falcon Landing sites, which have extensive Middle Archaic components tied to plant processing on the bajada, have Late Archaic period components (Hackbarth 2001; Wegener and Hall 2014), suggesting some continuity in land-use strategies through time.

Hall (in press) has reported on the Finch Camp site on the bajada in the southeastern Phoenix Basin that had a sizeable Cienega phase component dating to 300 BC. This component had multiple structures, many pits, and burials, along with early plainware ceramics (Wegener and Ciolek-Torrello 2011). In spite of its feature and artifact similarities to Cienega phase sites in the Tucson Basin, no maize was found, and the subsistence base appears to have been focused on mesquite, cactus, cheno-ams, and hunting. It is possible that some groups did not adopt farming, despite its prevalence in other portions of the region by 300 BC; this would have created an interesting dynamic between farmers who were looking for arable land and those who remained foragers, especially as population density increased. As with the Albuquerque Basin, where Railey (2016) suggests the interplay between farmers and foragers is something that should be investigated with further work, this transition may have resulted in social interactions that have implications for our understanding of subsequent Hohokam developments. By the Early Ceramic period (Red Mountain phase), agricultural villages were present throughout the Phoenix Basin, so additional research is needed to determine if this is the result of movement into the area by farmers or the further development of earlier farming communities that remain deeply buried in the river alluvium.

San Pedro River Valley

Eddy's work in the Cienega Creek Basin of southeastern Arizona was the first to document the extensive nature of Late Archaic occupations in this area (Eddy 1958; Eddy and Cooley 1983). The sites are located along the San Pedro River in a lush grassland setting with access to riparian resources including mesquite, cottonwood, oak, yucca, agave, and walnut. The presence of a cienega likely structured the Archaic period use of this area, as it would have provided cattails and other riparian resources along with suitable agricultural land. Eddy and Cooley (1983) noted that some of the sites they identified were buried by up to five meters of alluvium, so issues associated with alluvial burial in the Tucson and Phoenix basins are apparent here as well. No sites with evidence of early maize agriculture have been found in the San Pedro Valley, but this is most likely tied to site visibility factors rather than absence.

At the Donaldson (AZ EE:2:30) site (figure 5.10), Eddy identified a thick (sixty-centimeter) midden along the west bank of Matty Wash that he saw as evidence of prolonged occupation. Test excavations within the midden revealed multiple pits, including two shallow pithouse depressions with subfloor pits, hearths, roasting pits, bell-shaped storage pits, and flexed burials associated with San Pedro and Cienega points. Maize pollen was found in the midden, leading Eddy (Eddy and Cooley 1983:46) to propose that maize was a component of the diet of the San Pedro Cochise, with groups planting maize in fields bordering cienegas. These findings were the first to indicate that substantial changes occurred during the Late Archaic period in terms of population size, subsistence strategies, and sedentism.

The intensity of occupation documented at the Donaldson site led Huckell to continue excavations in the area in the 1980s as part of his research on early agriculture in the Southwest. Huckell (1995) excavated at both the Donaldson and Los Ojitos (AZ AA:2:137) sites (figure 5.10) and uncovered additional pithouses, storage pits, burials, a midden, and a large quantity of charred maize in deposits dating to 500 BC (Cienega phase). Artifacts from these sites expanded the known artifacts associated with early farming sites in the area to include trough metates, stone donuts and stone trays, along with shell beads and bracelets. As a result of these excavations, Huckell (1995) named the Cienega point as diagnostic of the latest stage of the Archaic period.

Flotation samples from both the Donaldson and Los Ojitos sites yielded a diverse array of plant remains, with maize and cheno-ams found in all samples and acorns found in all but one. Other wild plants indicate wide-ranging foraging activities, as agave, yucca, juniper seeds, saguaro, prickly pear, walnut, mesquite, and purslane were identified, along with squash or wild gourd. Faunal remains were dominated by large game, including mule deer, bighorn sheep, and pronghorn, indicating that game were still abundant in the area.

The increase in sedentism and technological changes that appear at these sites by 800–500 BC suggest an increased commitment to agriculture; this is seen in the capacity of storage pits, the presence of burials, the ubiquity of maize, and heavy teeth wear observed in the burials, suggesting a diet comprising a considerable amount of ground maize (Huckell 1995). These changes appear to be related to the lush environment of the Cienega Valley. As with the Phoenix Basin, it is not clear if these sites represent an expansion of farmers into the San Pedro Valley or *in situ* development from earlier Middle Archaic and San Pedro phase farmer-foragers, but the latter seems to be the most likely, given the lushness of this environment and the presence of Middle Archaic period sites throughout this area, akin to what has been found in the Tucson Basin.

Southeastern Arizona

Moving east out of the Phoenix Basin, several Cienega phase sites have been found along the Gila River in the Safford Valley during data recovery projects that have pithouses, storage pits, and artifact assemblages similar to those found in the Tucson Basin and San Pedro River Valley (Jones and Montgomery 2014). Middle Archaic sites have also been recorded throughout the Safford Valley and adjacent uplands. As with the San Pedro Valley, it is not clear if these Cienega phase sites represent the *in situ* development of farming communities like that observed in the Tucson Basin or the expansion of farmers into well-watered valleys.

A very different pattern associated with early maize is present further to the east in southeastern Arizona. Gilman's long-term work in the San Simon region of the Chihuahuan Desert has provided an extensive database on Archaic period hunter-gatherers and Ceramic period farmers in the region (figure 5.4). With the exception of the Gila River on the northern boundary, the environment is characterized primarily by dispersed springs and ephemeral streams that structured land use throughout the occupational sequence. Gilman et al. (2013) note that this region has less evidence of the use of maize during the Late Archaic period than surrounding areas and note that only one possible maize cupule has been found associated with Late Archaic period deposits. This lack of evidence of maize cultivation is interesting in light of evidence of Late Archaic occupations throughout the area (Gilman 1997). It appears that this area was used extensively by Middle and Late Archaic period foragers but that the focus was on hunting and gathering wild foods rather than on agriculture. This occurred despite the knowledge of agriculture, as maize cultivation was present in adjacent areas, so forager movements and social interaction would have led to the awareness of cultivation tactics in these other areas (Beale 2014). The lack of evidence of maize use during the Late Archaic is also interesting in light of the presence of maize cultivation in later Ceramic period occupations (Gilman 1997). This is therefore an example of foragers making a decision to remain focused on wild resources rather than to adopt cultigens. This is likely tied to the lack of reliable water sources available for planting and the subsequent risks involved in agriculture (Gilman et al. 2013). It may also be the result of the seasonal productivity and dispersed distribution of wild resources in the region, essentially selecting for a strategy of seasonal mobility.

Evidence of the Late Archaic period occupation of the Willcox Basin is more prevalent than the previous Middle Archaic occupation, possibly due to the refilling of the lake at 2000–1000 BC and the recharging of local springs (Waters and Woosley 1990). The number of sites increased significantly (from five to sixty-five), and site types include both camps and specialized procurement sites, which are primarily located in the foothills. Some base camps contain many features, including hearths and possible pits, and these are generally found adjacent to springs and on dunes next to the playa. These sites represent intensive seasonal occupation of the area by Late Archaic period foragers; however, like the San Simon region, no evidence of maize agriculture has been found.

Southwestern New Mexico

MIMBRES REGION

The earliest evidence of maize agriculture in the Mimbres region of southwestern New Mexico dates to the Cienega phase at the Forest Home (LA 78089) and Wood Canyon (LA 99631) sites in the Burro Mountains near Deming (figure 5.10; Turnbow 2000). Both sites had structures, bell-shaped storage pits, and burials dating to 800 BC–AD 300. One storage pit at the Wood Canyon site had maize phytoliths that dated to 800 BC, representing the earliest maize remains found in the Mimbres region. Abundant maize and squash remains were recovered from the Forest Home site, and maize was found stored in subfloor pits along with piñon, walnuts, and acorns. The earliest maize from this site dated to 385 BC, and Turnbow (2000) argues that it was repeatedly occupied over many years.

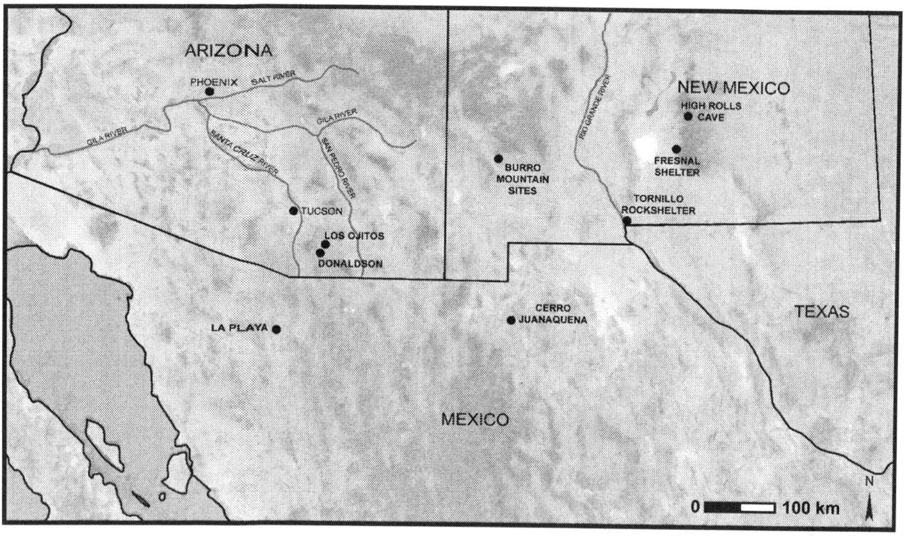

Figure 5.10. Late Archaic sites in southeastern Arizona and northern Mexico. Figure by Ben Van Alstyne

JORNADA REGION

The Jornada region of southern New Mexico has yielded evidence of early maize but does not show any evidence of a focus on maize in the diet or settlement changes associated with its introduction until substantially later in the Ceramic period, circa AD 1000 (Whalen 1994; Miller and Kenmotsu 2004). Most of the early maize dates have come from rock shelters, but as Miller (in press) has noted, this is due to the fact that subsistence studies have been most intensively conducted at rock shelter sites, so it does not necessarily represent a true pattern. Given the presence of early maize at open-air sites in other portions of the Southwest, open-air sites with early maize are probably underrepresented in this sample; this is supported by the recovery of maize pollen at LA 170280 discussed earlier.

The earliest well-dated maize from this region, not including the maize pollen from LA 170280, comes from Fresnal Shelter (LA 10101) in the Sacramento Mountains near the Tularosa Basin (figure 5.10; Tagg 1996). The shelter is located near a spring-fed stream, Fresnal Creek, in the piñon-juniper zone at an elevation of 6,270 feet. The presence of a reliable water source may explain its long-term use as a seasonal camp. Excavations conducted there in the late 1960s and early 1970s yielded multiple pits, hearths, and one burial, along with a dense, diverse artifact assemblage, including many perishables. The deposits span the time frame from 5900 to 300 BC, thus encompassing both the Middle and Late Archaic periods (Tagg 1996). Maize was prevalent in the deposits, and beans were also found, along with a variety of wild plants, including amaranth, ricegrass, and cacti. Two dates on maize place its use at Fresnal Shelter at the beginning of the Late Archaic period between 1200 and 1000 BC (Tagg 1996), and three dates on beans range from 170 BC to AD 100 (Table 4.2), supporting other data indicating that beans were first used much later than maize in the Southwest. As with rock shelter sites reported from the Mogollon Highlands, the initial use of maize at Fresnal Shelter appears to have been associated with a broad-based seasonal foraging strategy.

Early maize has also been found at High Rolls Cave (LA 114103), located in the Sacramento Mountains near Fresnal Shelter (figure 5.10; Lentz 2006). The cave has evidence of Middle and Late Archaic period occupations dating from 1500 BC to AD 300, with the main occupation dating from 1300–950 BC. The remains include storage pits, hearths, and perishables associated with Pinto, Chiracahua, Gypsum/Augustin, and San Pedro projectile points. Lentz (2006) argued that hunter-gatherers occupying High Rolls Cave emphasized hunting activities, given the large quantity of faunal bone and number of projectile points recovered from cave deposits. The earliest maize at the site dates to 1400–1000 BC. The site location provides access to a number of different vegetation zones, including desert resources, the piñon-juniper zone, and riparian vegetation along Fresnal Creek, and this is reflected in the diverse array of botanical remains, including maize, amaranth, acorns, piñon, dropseed, yucca, and agave. As with Fresnal Shelter, maize use at this site was associated with a seasonal foraging strategy.

Early maize has also been recovered from Tornillo Rock Shelter (LA 17687) located in the Organ Mountains north of El Paso (Upham et al. 1987). Ten maize cobs were found in excavated deposits at the site, and a composite sample of eight cobs yielded a date of 1700–1100 BC. The dating of the maize has been questioned because of the use of a composite sample, which may have incorporated younger and older maize. No projectile points were reported from the deposits and no other subsistence data are available to determine the nature of the occupation associated with this early maize use.

Hyland et al. (2003) reported on a study of perishable materials from cave and rock shelter sites in the Jornada area and used these data to argue for a migration of farmers into the region at 1200 BC. They argue that a suite of perishables, including coiled baskets, twill plaiting, S-twist cordage, and wickerwork sandals appear in the region between 1200 and 800 BC along with the earliest cultigens (although this would not be the case if the 2000 BC pollen dates are correct). These new forms co-occur with older textile forms, however, so there was no replacement of textile technology. Although Hyland et al. (2003) interpret the addition of new textile forms as indicating a migration of maize-using groups into the region, the co-occurrence of textiles is also consistent with the idea of local adoption, which also fits well with the early dates on maize. Beckes and Adovasio (1982) found that Archaic period textiles in the Jornada region showed close ties with northern Mexico, and these preexisting social connections would have facilitated the adoption of maize from groups in northern Mexico. It is also possible that ties with farmers to the west facilitated its adoption; our knowledge of the processes involved in this adoption will increase as additional work is done throughout southern Arizona, New Mexico, and northern Mexico.

The remainder of the Late Archaic period record, identified from extensive surveys in the Jornada area, indicates that groups expanded their resource use to all ecological zones in the region, perhaps as a result of wetter environmental conditions during this period (Miller and Kenmotsu 2004). Overall, the Late Archaic occupation appears to represent a land-use strategy characterized by seasonal movement and periodic aggregation. Miller and Kenmotsu (2004:227) argue that the use of maize during the Late Archaic was a component of diet diversification and that maize was used as a dietary supplement throughout this period.

Burned rock middens continued to be used for agave and yucca processing, and their use intensified during the Late Archaic period (Miller in press). This may indicate that groups chose to focus their efforts on these wild resources rather than intensify maize production, perhaps because it was a less risky strategy than maize agriculture. Miller (in press) suggests that these burned rock middens may have been an important component of ritual landscape use and served as loci where groups aggregated for social activities.

Maize use in the Jornada region increased during the terminal portion of the Archaic period (300 BC–AD 200) (Miller and Kenmotsu 2004; Miller in press). This was associated with the first use of bell-shaped storage pits, an increase in maize ubiquity at lowland sites, more formal site layouts, and more trash deposits. Miller (in press) suggests that sites with these features, much like Cienega phase sites to the west, document increased sedentism, population growth, and more maize dependence, and were the precursors to later Mesilla phase (Ceramic period) occupations.

Northern Mexico

LA PLAYA

La Playa (SON F:10:3) is a massive multicomponent site located on the floodplain of the Rio Boquillas in northern Sonora (figure 5.10). The site covers an area of twelve square kilometers and contains evidence of occupation from the Paleoindian through the Spanish Colonial periods (Carpenter et al. 2005). A report on the excavations that have been conducted there over the last decade has not been published, but some information on its occupation is available (Carpenter et al. 2002, 2005; Mabry et al. 2008; Watson 2007).

The site is located near a marsh habitat, and the river provided a year-round water supply, so the abundant resources may account for the intensity and temporal span of the occupation. Many other Archaic period sites are present in the region, including shell midden sites along the coast and trails and intaglios in the desert (Carpenter et al. in press) so the Archaic period occupation may be one component of a regional settlement system. Carpenter et al. (1999) argue that the major occupation of the site dates to the Late Archaic period, from 1500 BC to AD 200, based on radiocarbon dates from roasting pits and burials and on the prevalence of Late Archaic period San Pedro and Cienega projectile points. The Middle Archaic component is represented primarily by Pinto/San Jose, Chiricahua, and Cortaro points, and it is likely that some of the many features that are present on the site also date to this phase.

The site has thousands of features, including over one thousand roasting pits, bell-shaped storage pits, over two hundred burials, most of which date to the San Pedro and Cienega phases, many cremation burials, dog burials, and numerous post-Archaic earthen mounds (Carpenter et al. 2005; Watson 2007). Maize is ubiquitous in all of the roasting pit features investigated thus far. Associated wild plant remains include amaranth, cheopodium, cholla, prickly pear, and mesquite. Despite the large size of the site, the number of roasting pits and burials, and the dense, diverse artifact scatter, no Late Archaic period structures have been reported. The site is heavily eroded, so this could be a function of geological processes, but it also may indicate that this site represents a unique land-use strategy.

The artifact assemblage includes abundant lithics and ground stone, along with more specialized artifacts such as eared stone trays, stone cruciforms, and tabular tools. Evidence for shell ornament production has also been found, possibly beginning as early as the Late Archaic period (Carpenter et al. 2005).

The extent of the site makes it unique in the southern Southwest as it is substantially larger than even the largest Cienega phase sites in the Tucson Basin, such as Los Pozos (Gregory 2001) and Santa Cruz Bend (Mabry 1998a). Although Carpenter et al. (2002) use data from La Playa to argue for a migration of farmers into the Southwest, the data do not currently support that interpretation because evidence for maize at the site postdates the arrival of maize in adjacent southern Arizona. It is possible that earlier maize will be found at La Playa, but it will be important to investigate the nature of the occupation associated with its presence to contextualize the nature of the occupation associated with it. If, as will be argued in chapter 6, cienega environments served as a major loci for early maize farming from which it spread via new and existing social networks, then La Playa would be expected to have played an important role in this process.

CERRO JUANAQUEÑA

Perhaps the biggest challenge to existing notions of early agriculture in the Southwest comes from work at the site of Cerro Juanaqueña, located on an isolated hilltop along the Rio Casa Grande in northwest Chihuahua (figure 5.10). This is a terraced basalt hillside (*cerro de trincheras*) with evidence of extensive labor investment in the construction of terraces for habitation (Hard and Roney 1998, 2005; Roney and Hard 2002). Two episodes of occupation have been identified, with the earliest dating to 1250–1100 BC and the second from 400–100 BC. These postdate the earliest maize agriculture in other portions of the Southwest, and we do not yet know the circumstances surrounding the initial use of maize in northern Mexico.

A total of 550 terraces have been mapped at Cerro Juanaqueña (Hard and Roney 1998, 2004). Rock rings found within the terraces are thought to represent house foundations that contained brush structures; they are oval and average two meters in diameter. These features are surrounded by what Roney and Hard (2002) refer to as "macrofeatures" or "perimeter berms" that define the exterior boundaries of some of the terraced areas. Roney and Hard (2002:174) estimate that terrace construction required extensive communal labor, in the range of 30 person-years, and argue that the site may have housed as many as two hundred people, indicating a population density and degree of sedentism unparalleled in other portions of the Southwest at this time. These interpretations have been debated by some researchers, however, who argue for a less intensive nature of the occupation (Vierra 2005; Schmidt 2008).

The artifact assemblage from Cerro Juanaqueña includes abundant chipped and ground stone tools, including San Pedro points, a variety of bifacial and unifacial tools, many manos, heavily worn metates, mortars, pestles, stone pipes, cruciforms, and spindle whorls. Some of the metates were worn to the point of no longer being usable. Vierra's (2005) study of the lithic assemblage indicated it was similar to other Late Archaic residential sites in the Tucson Basin, but exhibited more bifacial technology than sites on the Santa Cruz River, reflecting a range of core reduction strategies using primarily local river cobbles.

Maize is ubiquitous in flotation samples from the site. A range of wild plant resources were used as well, including cacti, mesquite, walnut, and grasses. Faunal remains are dominated by rabbits, but include a fish bone and bison bone, and Hard and Roney (2004:289) suggest that fish may have been exploited relatively intensively. Although it was first reported that domesticated amaranth was present at Cerro Juanaqueña (Roney and Hard 2002:165), this has not been supported in subsequent analysis (Browman et al. 2005). Schmidt's (2008) analysis of the faunal remains indicates that these *trincheras* sites in northern Chihuahua, although large and spectacular, were seasonally occupied. Schmidt argued that groups used the terraces for agriculture during the summer but moved in other seasons to hunt and gather other resources. This interpretation contrasts with that of Hard and Roney (2005), who argue that the occupants of Cerro Juanaqueña were highly agricultural and sedentary.

Hard and Roney (2007) suggest that the episodic dates of occupation indicate that the site was used primarily for defensive purposes, as groups moved onto the hilltop sites during times of conflict. They thus view this site as a short-term strategy in response to regional conflict. They argue that several other *cerro de trincheras* occupations located during surveys of the surrounding area served similar functions as thirteen *trincheras* sites with similar characteristics have been found in the Rio Casas Grandes watershed. Unfortunately, little work has been done on the Archaic period occupation of this region outside of the surveys conducted by Hard, Roney, and MacWilliams, so it is not clear whom groups were fleeing from or what the conflict was about, if, in fact, these are defensive sites. It is also possible that they represent seasonal aggregation sites, perhaps for social and ritual purposes, akin to the ritual aggregation sites identified by Miller (in press) in the Jornada region of southern New Mexico. This would also fit with the artifact assemblage, including the recovery of specialized materials such as pipes and trays, but it does not explain the labor invested in the construction of the terraces. Currently, it is difficult to place these sites into context in terms of Late Archaic period developments because they are unique in the region.

Trincheras *Sites in Southern Arizona*

Several *trincheras* occupations have been found in southern Arizona, apparently adjacent to local farming populations. Maize dating to 1200 BC was found on the *trincheras* site of Tumamoc Hill in the Tucson Basin (Fish et al. 1986); this site was initially thought to date to the Hohokam Classic period, but further investigations have documented a Cienega phase component with many terraces (Fish et al. 2011). Extensive mapping of the features has revealed a circular layout and Fish et al. (2011:188) view several massive linear features on the site as representing "the earliest known public architecture in southern Arizona and possibly in the U.S. Southwest." Excavations on the summit of Tumamoc Hill revealed a communal structure similar to those found at other Cienega phase floodplain sites. The characteristics of this Cienega phase *trincheras* site lend some support to the idea that these sites were perhaps important ritual sites on the landscape.

Roney et al. (2014) reported on the Round Mountain site, a *trincheras* site located on the floodplain of the Gila River, which, as with Tumamoc Hill in the Tucson Basin, currently represents an isolated occurrence of this site type in southeastern Arizona. Preliminary fieldwork has documented many terraces and rock rings that are similar to those identified at *trincheras* sites in Chihuahua. Like Cerro Juanaqueña, a large quantity of ground stone is present, including stone pipes. Test excavations indicate the occupation dates to the Cienega phase, circa 500–400 BC (Roney et al. 2014). Additional work is planned for this site, which should help clarify its role as an early farming settlement in this region.

It is as yet unclear what these occupations in southern Arizona represent. They could represent groups moving into the region from northern Chihuahua in search of arable land or fleeing conflict. It is also possible that they represent the borrowing of the strategy of building *trincheras* for defense or social gatherings by local farmers. This seems unlikely given the close similarities seen in the features at these sites and the lack of similar features at any other sites in southern Arizona. It is also possible that these sites represent farmers from northern Mexico who migrated into the Gila River Valley and the Tucson Basin. In this case, there were already farmers in the major river valleys, however, which may have made these social interactions more complicated.

Trincheras *Sites and the Late Archaic*

Cerro Juanaqueña and other Late Archaic period *trincheras* sites represent a challenge to researchers examining agricultural beginnings because they differ so much from the other sites with early agriculture in the Southwest, yet contain San Pedro points, stone tools, and ground stone characteristic of southwestern Late Archaic sites (Beale 2014; Vierra 2005). The presence of San Pedro points has been used to argue that Cerro Juanaqueña is part of the Southwest Late Archaic period; however, as Roth (2014) has discussed, the other archaeological data, especially the architecture and site characteristics, suggest that it may be culturally distinct from other populations in the southern Southwest. No burials have been found at the site, so this is a tentative assumption, but the divergence of this occupation from other Southwestern populations calls into question whether it represents a related group. Perhaps a consideration of social networks and identity formation will aid in the assessment of these *trincheras* sites. It is possible that these groups were identifying with northern Mexican groups more than with the Southwest, and we are misclassifying them as Southwestern because of the presence of similar projectile point styles. The role of identity formation in the Archaic period is understudied, but it may have had an important impact on their behavior. If these are separate ethnic groups, then this may explain the lack of congruity with other Southwestern data and the presence of distinct traits such as terraced hillsides and possible conflict found at these sites. Clearly the answers are not available with current data. We need more information on the Archaic period predecessors to these groups, more data on these occupations, and more data on the interactions with other groups in southern Arizona to assess the nature of these occupations.

The Role of Irrigation in Agricultural Intensification

Mabry (2008) has discussed in detail the multiple techniques used for doing agriculture in the Southwest, and irrigation remains a hallmark of Late Archaic groups in southern Arizona. Both Mabry (2005b) and Thiel and Mabry (2006b) note that there are no known precedents for these early irrigation canals in Mexico and argue that irrigation was a local innovation in southern Arizona, with the earliest canal dating to 1500 BC at the Clearwater site. The role of irrigation networks in both agricultural beginnings and in the spread and integration of agriculture appears to have been important, at least in riverine settings in the Tucson Basin. Mabry (2008) estimated the labor involved in the construction and use of the canals at Las Capas, arguing that initial labor may not have interfered with subsistence activities and other tasks, but that regular cleaning and periodic rebuilding "entailed a considerable investment of time and labor" (Mabry 2008:182).

At a minimum, the use of this technique would have increased productivity and minimized the risk associated with maize agriculture, and potentially could have resulted in the production of surpluses. Irrigation is the major feature that distinguishes sites in the Tucson Basin from those elsewhere, and the presence of irrigation networks has been one of the key factors used to argue for long-term sedentism and more complex social behaviors for the earliest maize farmers in this portion of the Southwest (Mabry 2002).

These canal systems may represent a threshold point where increased sedentism was required for their maintenance, and the additional labor invested in irrigation provided increased food productivity and the ability to support higher population densities. They may have provided the basis for some of the changes observed during the Cienega phase, as population increase led to the need for more formal storage, more formal structures, exchange networks to buffer risk, and social integration mechanisms such as the use of communal structures.

The question of why irrigation was developed in this area has not been addressed. Archaeological evidence from the Tucson Basin points to the presence of Middle Archaic hunter-gatherers on the floodplain, and based on data on the earliest maize from the Clearwater, Las Capas, and Los Pozos sites, maize was integrated into seasonal gathering activities (Roth and Freeman 2008). The earliest canals at Clearwater and Las Capas are not associated with any other evidence of agricultural intensification, but at Las Capas, the San Pedro phase canals are associated with increased sedentism (Mabry 2008; Vint and Nials 2015). No evidence of subsistence or environmental stress is associated with the innovation of irrigation at 1500 BC, and in fact it appears to have occurred during a wet period when resources were abundant and floodwater and water-table farming would have been viable strategies. The extent of the San Pedro phase irrigation system illustrates that labor was being invested in this new technique, but macrobotanical data do not indicate that irrigation resulted in dramatic subsistence change or in changes in the maize itself (Diehl 2005a,b, 2015). It may be that the environment allowed experimentation with new techniques that enhanced productivity and were therefore further developed.

Summary

Current data indicate that the initial adoption of maize in the southern Southwest occurred primarily in well-watered floodplain settings and that maize was added as a component of a seasonal foraging strategy. All of the early sites have evidence of hunting and gathering along with maize agriculture, and all exhibit some level of residential mobility, although only small portions of the deeply buried floodplain sites have been excavated.

When Huckell (1990, 1995) first suggested that farmers entered southern Arizona via a migration from Mexico along the major river valleys, he based this inference in large part on a comparison of the known data on the Middle Archaic period with data gleaned from excavations at Late Archaic floodplain sites conducted in the late 1980s and early 1990s. The growing evidence from sites with maize dating to 2100 BC indicates that maize agriculture in this area predates the site characteristics described by Huckell by up to one thousand years. As with the northern Southwest, we need to separate our understanding of the initial use of maize in the southern Southwest from what happened after agriculture was integrated into the diet. Current data on the earliest maize are more parsimonious with what would be expected as local hunter-gatherers adopted maize agriculture and associated storage technologies from groups that they were in contact with in northern Mexico. It is thus time to rethink some of our models of agricultural origins in this region in light of these data.

It is apparent from examining the database on the Late Archaic period in the southern Southwest that much like the northern Southwest, there is variability in the response by hunter-gatherers to the introduction of maize. Some groups, especially those along the major waterways in southern Arizona, intensified their production of maize; although this did not result in genetic changes in maize or production technology, it did result in changes in settlement strategies, as groups begin to construct more durable houses and large-capacity storage pits, suggesting that they were spending more time at floodplain settlements. In the Tucson Basin, this coincided with a commitment of labor to irrigation. Changes in the artifact assemblage are also apparent, with the addition of new forms of ground stone, ceramic figurines, and incipient plain wares and different forms of shell jewelry. These changes occur over a time span from 1200 BC to AD 200, but ultimately set the stage for the developments that occur during the Ceramic period.

The impact of the adoption of maize in other portions of the Southwest was more muted. In some areas, such as the San Simon region of southeastern Arizona, maize was not adopted by Archaic period hunter-gatherers and was not used until later in the Ceramic period. In areas like the Jornada region of southwestern New Mexico, maize use occurred early, perhaps by 2000 BC along playa margins, and is present in rock shelter sites by 1200 BC as part of a seasonal foraging strategy much like that observed at rock shelter sites in the Mogollon Highlands. This "low-level food production" continued through the Ceramic period in this region.

The earliest evidence for maize agriculture in some of the areas where early maize use would be expected—the Phoenix Basin, San Pedro River Valley, and Gila River Valley—comes from Cienega phase sites postdating 800 BC. It is not clear if this is a result of the movement of already established farmers into these settings or if earlier evidence of maize use is present but buried beneath the alluvium. Only continued work in these regions focusing on deeply buried deposits will allow us to determine the nature of agricultural beginnings in these areas.

Finally, two sites in northern Mexico, La Playa and Cerro Juanaqueña, have characteristics that differ substantially from those found elsewhere in the southern Southwest—La Playa in terms of its size, number of burials, and lack of structures, and Cerro Juanaqueña in terms of its terraces and rock structures with evidence of considerable labor investment. These sites again postdate the initial adoption of maize, dating to 1200 BC and later, and the circumstances surrounding the initial use of agriculture in both areas are not currently understood. Additional work is therefore needed to determine if these represent unique situations or separate groups with distinct identities or are part of the variability represented in Late Archaic occupations across the southern Southwest.

Exploring Agricultural Beginnings 6

There is also, very clearly, no single and straight developmental pathway across this middle ground, no single evolutionary track leading to agriculture.

—SMITH 2001b:301

AGRICULTURAL BEGINNINGS IN THE SOUTHWEST can be categorized by one word—variability. This book has focused on the variability that can be seen in agricultural beginnings across the northern and southern Southwest in the way that maize agriculture was initially introduced and the ways in which it was integrated into the subsistence regime of prehistoric hunter-gatherers. This variability encompasses the environmental setting, which ranges from the Basin and Range desert settings of southern Arizona and southwestern New Mexico, to the piñon-juniper uplands of the Mogollon Rim and the Rio Grande Valley, to the mesas and canyons of northern Arizona, southwestern Colorado, and southeastern Utah. It also includes the varying nature of the Middle Archaic period occupations that preceded the introduction of agriculture and variation in the subsequent impact of the adoption of agriculture on the foragers who began to farm.

Yet within this variability, certain patterns are present, and these can help us begin to understand the cultural processes associated with the transition from hunting and gathering to farming. Similarities are seen in the kinds of settings in which the earliest agriculture was practiced, especially in terms of the availability of reliable water, arable land, and wild resources, the timing of the earliest agriculture across vast portions of the Southwest, and the nature of the earliest agricultural occupations, all of which appear to represent the integration of maize into seasonal foraging activities. We can use both the similarities and differences to explore the factors involved in the adoption of agriculture by prehistoric hunter-gatherers in the Southwest, including the factors influencing the initial decision to adopt and experiment with maize and later decisions on what to do with this new subsistence item in terms of the degree of integration into subsistence strategies and whether to intensify production or not.

This chapter summarizes the similarities and differences that we see in agricultural beginnings in the Southwest, explores some of the factors that influenced these patterns, and examines how this relates to what we know about the behavior of ethnographic hunter-gatherers. Ideas for future research as we move forward in understanding this important transition are then presented.

What We Know about Agricultural Beginnings

One of the major points coming out of this book is that we must decouple studies of the initial adoption of agriculture from what occurred after agriculture was introduced to the Southwest. As discussed in chapter 2, the migration model was originally developed based on Late Archaic and Basketmaker sites that substantially postdate the initial introduction of maize agriculture to the region. It is thus important to separate the mechanisms and processes associated with the initial use of maize at 2100 BC from the adaptations that we see after 1000 BC. These later occupations cannot be viewed as resulting from the same kinds of decisions and cultural processes as initial maize use, as many hunter-gatherers had already made the transition to being farmer-foragers by this time. By separating the two time periods and associated cultural traits, it is possible to develop more coherent and perhaps more accurate models of both the initial use of maize in the Southwest and the variability in the subsequent impact of maize on these groups.

It is now apparent from multiple sites in a variety of environmental settings across the Southwest that maize was initially adopted by Middle Archaic period hunter-gatherers as part of a seasonal foraging adaptation by 2100 BC (figure 6.1). These sites and site data indicate that maize and the knowledge of how to grow and store it spread relatively quickly, probably via existing kin and interaction networks like those proposed by Merrill et al. (2009). The database that is coming out of work in both the northern and southern Southwest thus strongly supports the model that indigenous hunter-gatherers adopted maize from other early farmers. Despite the wide-

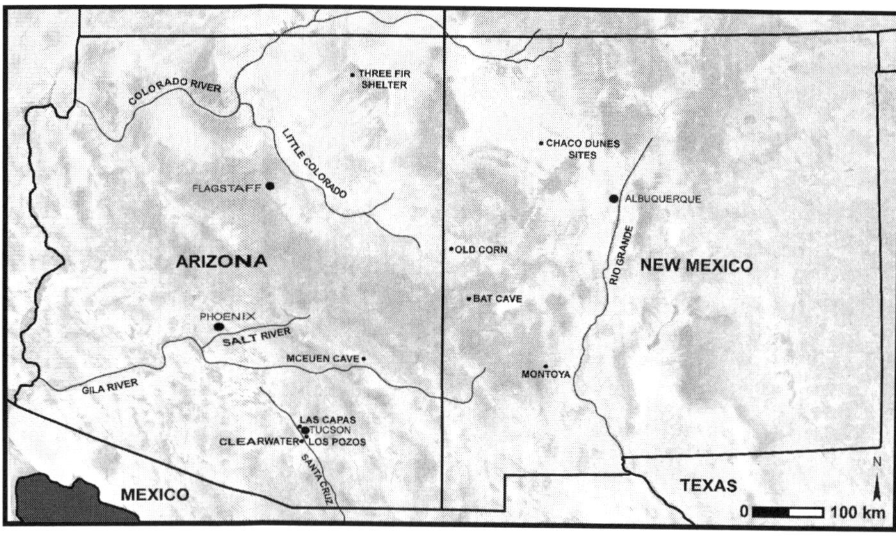

Figure 6.1. Earliest sites with maize in the Southwest. Figure by Aaron Woods

spread occurrence of early maize in settings throughout the Southwest, not all hunter-gatherers in the region initially adopted maize, and the integration of maize varied depending on local circumstances, including environmental setting, wild-resource availability, and local population density.

Many of the earliest sites in both the northern and southern Southwest are in well-watered alluvial settings where summer gathering efforts may have facilitated the integration of maize. These include dune sites in the San Juan Basin, sites on alluvial terraces in southern and central New Mexico and northeastern Arizona, and floodplain sites in the Tucson Basin. The fact that the initial use of maize is associated with the beginning of a wet period between 2500 and 2000 BC is undoubtedly important, as this would have provided a more secure environment in which to experiment with this new cultigen, especially after the climatic perturbations witnessed during the previous Middle Holocene. This wetter climatic cycle would have recharged springs and made even seasonally available water resources more reliable. Certain wild resources such as cheno-ams and seasonal grasses would have been more abundant, and would have supplemented the already existing staples such as piñon in the north and mesquite and cacti in the south. The increased density of wild plants would also have provided for and attracted more game. These resource-rich settings would have enabled Middle Archaic period foragers to afford the risk-taking that would be associated with maize agriculture, as they ensured an adequate diet even if the maize crops failed. Cienega environments and dunes appear to have been particularly targeted locales, perhaps because of the availability of a diverse array of wild resources. Cienega and alluvial terrace settings would have enabled farming using water-table techniques. Playas in southern New Mexico may have served as similarly suited locales.

It appears that in all cases in the Southwest, maize was originally adopted as a dietary supplement (see Minnis 1985, 1992), yet one with enhanced storability and the potential for manipulating productivity compared to wild plants. This pattern is consistent with what would be expected as hunter-gatherers adopted a new food source. As noted in chapter 3, foragers are generally risk averse, so it is unlikely that they would adopt a new subsistence item unless it provided something that made it worthwhile to do so. Early maize was not highly productive, but it provided a degree of predictability that was not present in all of the higher-ranking wild resources and could be manipulated to increase productivity, which was not a characteristic of wild plants. As a grass, its growth and processing requirements were similar to those of the wild grasses that hunter-gatherers were already exploiting, and this likely had an influence on its adoption, as it did not require new techniques for processing, nor did it require a substantial change in land use, given that it could be grown in the same areas where wild resources were available. The major technological change associated with the adoption of maize was storage, and most researchers think that the storability of maize and the added benefit of being able to store wild resources, was a key factor in its adoption (Dello-Russo 2006; Hunter-Anderson 1986). Storability may have been more important in upland and northern settings where it enabled overwintering strategies to develop (Hunter-Anderson 1986; Smiley 2002), which probably accounts for the recovery of early maize at rock shelter sites in the northern Southwest. However, the presence of bell-shaped storage pits at all of the earliest agricultural sites in the southern Southwest attests to the important role of storage in this initial adoption and illustrates that storage technology was adopted with the maize.

This exemplifies another important aspect of agricultural origins, for as discussed in chapter 3, maize was not introduced into a vacuum but arrived to hunter-gatherers who were well versed in the challenges of their environment. They knew when the stress periods would be for both plant and animal availability, and this may have made maize, with its predictable and storable crop, particularly attractive. They knew enough about the plants that grew locally to know what the issues were that would be involved with growing crops and could ask questions when adopting maize to gain information on the best strategies for dealing with these issues.

The adoption of maize also illustrates the important role that social networks played in Archaic period lifeways as this process involved information exchange, and that information came through social interaction. Information that would have been passed along with the maize included techniques for producing a viable crop, solutions to some of the problems associated with doing agriculture in an arid environment, and storage techniques, which would ensure success. Although it is most likely that women were the primary people learning about these techniques, given their close association with plant gathering and processing, the initial adoption of agriculture would have been a group-level decision, as cross-cultural ethnographic data on hunter-gatherers show that major decisions impacting members of a hunter-gatherer society are made by group consensus.

The rapid spread of agriculture across the Southwest suggests that this exchange of new cultigens and ideas about them was not just accomplished via kin networks but also through larger social networks, perhaps akin to those that would have been necessary to ensure mates (e.g., McBrinn 2005). The first hunter-gatherer groups to adopt maize were probably those in the most productive settings for risk taking, with extensive social networks that could provide backup resources, or those in areas where agriculture was perceived as having a significant benefit. They then passed these new cultigens and techniques to relatives, friends, and trading partners, who in turn either integrated maize into their subsistence regimes or did not, depending on their particular circumstances. Intermarriage likely played an important role in both the initial adoption of maize and in the spread of maize to other forager groups, as marriage partners could introduce cultigens and agricultural techniques to a group. Some groups readily adopted maize agriculture; again this occurred primarily in settings where it was easily incorporated into existing subsistence strategies or where the nutrients, predictability, and storability were deemed as desirable characteristics. Others may have been more reticent to adopt until they were able to determine how agriculture worked for their neighbors; this may account for some of the lag time that we see from the initial adoption of maize at 2100 BC to the prevalence of maize-using groups throughout the Southwest by 1000–500 BC. In some cases, farmer-foragers moved into new areas, bringing maize agriculture with them, which may account for the appearance of agriculturally based Basketmaker groups in the northern Southwest.

Flexibility appears to have been a major characteristic associated with the earliest agricultural occupations and this, too, fits well with what is known about ethnographic hunter-gatherer behavior. This does not mean that maize was cultivated without effort; Wills (1992) has emphasized that even small-scale maize agriculture required certain labor investment in field clearing, planting, harvesting, and storing. Welch (1991) discussed the horticultural techniques practiced by the Apache and showed that groups integrated gathering, hunting, and farming through scheduling and the use of task groups. He also notes that weeds, insects, birds, mammals, and hail sometimes resulted in the loss of up to two-thirds of the crop, which was acceptable because of the subsistence contributions by wild plants and game. It appears that this same strategy was used by many Archaic period hunter-gatherers in the Southwest as they integrated maize into existing foraging regimes. This may also account for the prolonged period of low maize dependence after its initial introduction, akin to Smith's (2001b) low-level food production. Some Apache groups practiced ditch irrigation, leaving some individuals (very young or old) to weed and irrigate while the main group left on hunting and gathering forays. Thus, small-scale irrigation is also not incompatible with limited agricultural dependence, possibly accounting for the presence of canal irrigation at sites like Clearwater by 1500 BC.

Phillips (2009) has stressed the concept of marginal cost in examining the incorporation of agriculture into the subsistence base in an attempt to determine the circumstances that would make the increased labor effort invested into clearing, planting, and tending the fields worth doing. He notes that the best combination of wild and domesticated resources may vary from region to region, and this is what we see across the Southwest. Phillips argues that with low population densities and foraging with adequate returns, farming would remain a supplemental strategy. This appears to represent the earliest use of maize in the Southwest. According to Phillips (2009), with increasing populations, farming would be added as a subsistence strategy to provide a more reliable and productive food resource, and as population continued to increase, intensification would occur.

The Impact of Agriculture

In addition to the similarities and differences observed in agricultural beginnings across the Southwest, we see variability in the integration and impact of agriculture. During the Late Archaic and Basketmaker II periods, major differences start to develop between the southern and northern Southwest that can be tied directly to the different ways that maize agriculture was incorporated into the subsistence and settlement system. In upland settings like the Mogollon Highlands and Rio Grande region, maize remained a supplemental component of a foraging strategy well into the Late Archaic period, illustrating long-term stability in this strategy (Wills 1988a,b, 1992; Vierra 2008; Post 2013). On the Colorado Plateau in northern Arizona, southeastern Utah, and east into southwestern Colorado, with the exception of some rock-shelter and open-air sites with maize predating 1000 BC (see chapter 4), most of the occupation occurs with Basketmaker farmers, who, by all accounts arrive with fully developed agricultural adaptations. As discussed in chapter 4, this is likely due to the environment, which was very risky for agriculture. Groups may not have moved into this region until agriculture was a productive subsistence strategy providing storable calories for overwintering.

In some areas of the southern Southwest, groups exhibit similarities to northern groups in their incorporation of maize agriculture. In the Jornada region of southeastern Arizona, the same long-term stable strategy of mixed farming, gathering, and hunting that is seen in the Mogollon Highlands is present, extending in some areas into the AD 1000s (Mauldin 1996; Miller and Kenmotsu 2004). In other, more arid portions of the southern Southwest, such as southeastern and far western Arizona, agriculture was sometimes not present at all, and when it was, it occurred much later in the Ceramic period. This was largely a result of the environment, as regions lacking reliable water sources and patchy resource distributions would have been some of the least attractive for farming (Gilman 1997).

An entirely different pattern is present in the Tucson Basin along the Santa Cruz River floodplain and in the San Pedro River Valley, where Late Archaic sites with pithouses, large-capacity storage pits, hundreds of other pits, and burials exhibit a higher level of sedentism and eventually a higher level of agricultural dependence than is seen elsewhere. Despite the significantly different archaeological signature that these sites contain, macrobotanical, faunal, and chipped and ground stone data indicate the presence of a very stable, long-term forager-farmer adaptation (Adams 2005, 2015; Diehl 2005b,c; Dean 2005; Sliva 2005). The degree of sedentism represented at these sites is still a topic of debate, but there is no question that they represent a fundamentally different pattern than that seen at other sites in the northern and southern Southwest. The degree of sedentism may be more related to the practice of irrigation and land tenure than agricultural dependence per se, which makes them even more interesting in light of studies that have documented the correlation between sedentism and agricultural dependence (but see Rocek 1995, 1996). The two Late Archaic sites in northern Mexico discussed in chapter 5 highlight the fact that these early agricultural communities were linked to a larger regional network, possibly in complex ways that may have included conflict (Hard and Roney 2007).

This leads to the issue of identity formation and its role in the adoption and spread of agriculture. The broad social networks that existed during the Archaic period discussed by McBrinn (2005) and Beale (2014) facilitated interaction, exchange, and intermarriage on a scale necessary for hunter-gatherers living in small populations dispersed across the region. As Del Castillo et al. (2014) discuss, hunter-gatherer groups defined themselves as similar based on their degree of interaction. As foraging territories developed, population densities increased, and logistical mobility strategies became more widespread resulting in less contact with members of the larger social network, differences in cultural identity may have developed and been emphasized. These may be manifest in differences in architecture and material culture and may account for some of the differences that we see between groups in the northern and southern Southwest and especially in northern Mexico.

Moving Forward

Although we now have an emerging picture of agricultural beginnings in the American Southwest, there are still several areas where additional work can help enhance our understanding of the transition from hunting and gathering to farming. As noted in chapter 1, this book has focused on the strong cases where archaeological evidence for early agriculture and its associated cultural context is available. As cultural resource management and other research projects continue, this picture will be further enhanced and potentially modified. This is especially true for areas where little work has been done and for areas where the potential for deeply buried sites such as the Phoenix Basin and San Pedro River Valley may change our interpretations of how agriculture was introduced and integrated in those areas.

In this section, I have identified six areas where additional research can contribute to our knowledge of the adoption and integration of agriculture. These highlight the significant gains that can be made as we build on the research that has been done up to this point.

Interactions between Farmers and Foragers

Several researchers have noted that during the period when maize was adopted and integrated into the subsistence base, some groups remained hunter-gatherers while others made the shift to farming (Railey 2016; Vierra 2008). This topic is of critical importance if we are going to understand the nature and complexities of agricultural beginnings. The relationship and degree of interaction between early farmers and surrounding hunter-gatherers may provide information on how and why agriculture spread. One issue that should be addressed with continued research is whether conflict was involved in the expansion of farmers, as has been proposed for northern Chihuahua and possibly southern Arizona by Hard and Roney (2007). If conflict is present, then it should be represented in the archaeological record and burials at these early farming sites.

A second issue related to this topic involves the factors influencing farmer movements, especially into lush regions such as the Phoenix Basin, San Pedro River Valley, and Gila River Valley. Each of these areas has evidence of extensive Middle Archaic period hunter-gatherer occupations, but the earliest farming sites date to the Cienega phase, usually after 500 BC. Is this because of site visibility issues, with earlier farming sites present in deeply buried deposits like those found in the Tucson Basin? Or were farmers seeking prime farmland, displacing hunter-gatherers who occupied these river valleys? This can be answered through additional excavations at alluvial sites and with detailed comparisons of artifact assemblages to evaluate the degree of continuity represented in these assemblages.

Questions of Basketmaker developments and movements should also be addressed with future research to evaluate the relationship between Basketmaker farmers and Archaic period hunter-gatherers throughout the northern Colorado Plateau. Ideas about what Basketmaker is and how it developed are still debated. Extensive studies of identity formation have been done in other parts of the world (Amundsen-Meyer et al. 2011; Díaz-Andreu et al. 2005; Meskell 2001, 2002), and similar datasets could be used to evaluate identity formation in Basketmaker groups, like the technological studies conducted by Geib (2002).

Demography

One thing that becomes increasingly clear as we examine the archaeological database for the adoption of maize and the resulting consequences of cultivation is that we need better control of demography. More sophisticated demographic models are needed to address issues such as the relationship between population growth, population density, climate change, and the resource base to further determine the role of the environment and population density in the adoption of cultigens. This can be done by developing techniques for reconstructing demographic profiles. If population density had the impact that many researchers suggest that it had on early farmers, then it behooves us to explore ways to more accurately determine population size.

Innovation and the Integration of New Technologies

Although some aspects of behavior show continuity through time as hunter-gatherers made the transition to farming, technological change is apparent both at the initial adoption of agriculture and during subsequent periods of integration and intensification. Storage technology appears to be the major technological change associated with the first use of maize, as no changes in chipped or ground stone technology have been identified with this transition. Further studies of pit storage techniques and the issues associated with them may refine our understanding of the constraints imposed by pit storage and aid in determining if this was the reason that groups did not intensify their production beyond a certain point (Diehl and Waters 2006). Adams (2005, 2015) has done extensive studies of ground stone tools from Late Archaic farming sites in the Tucson Basin and has documented their role in food production and pigment processing; she has also established that they are not directly related to maize dependence, but instead are tied to overall food production strategies (Adams 1999). Further detailed studies like these of both ground and chipped stone tools may help us examine the factors that led to continuity in material culture and changes in technology that can then help determine when agriculture became a subsistence focus. Geib and Jolie (2008) have done detailed studies of basketry to establish that coiled baskets were first used during the Early Archaic period; these kinds of studies of Middle and Late Archaic period basketry may provide insights on both subsistence change and identity formation. Finally, Mabry (2002) has argued that irrigation agriculture was independently developed in the American Southwest and was present very early in the Tucson Basin, by 1500 BC. The role of irrigation technology at early farming sites and the lack of subsistence change documented by Diehl (2005b) should be investigated more thoroughly as we move forward, as it continues to represent an enigma in our understanding of the transition from foraging to farming.

Sustainability

Sustainability is a hot topic in our society today, and studies of agricultural beginnings can contribute to this global discussion. Different strategies for incorporating maize agriculture resulted in differing levels of dependence across the Southwest. Sites like Old Corn and the Chaco dunes exhibit the earliest evidence of maize agriculture in the Southwest, yet this strategy was not sustainable in the long term in these environments, as the early experimentation was not followed by any intensification or continued occupation. In other areas, like the Jornada and Rio Grande regions, a low-level food production strategy with a mix of foraging and supplemental farming was sustained for several thousand years until Ceramic period times. In well-watered alluvial settings like the Tucson Basin and San Pedro River Valley, large farming sites are present with many structures, storage pits, and more complex artifact assemblages associated with irrigation agriculture, documenting a different path toward agricultural intensification from that seen in other regions. In some portions of the northern Colorado Plateau, it does not appear that agriculture was practiced at all until Basketmaker farmers moved in with their sustainable overwintering strategy. By focusing on the mechanisms and processes responsible for these differences in sustainability, we can learn more about the factors influencing the initial decision to adopt agriculture and later intensify production.

The Role of Ritual and Ideology in the Transition

Research at Neolithic sites in the Near East points to a significant role of ritual in the transition from foraging to farming (Kuijt 1996, 2000; Verhoeven 2002). This is one of the most understudied aspects of agricultural beginnings in the Southwest, although several researchers have discussed the possibility that ritual played an important role in this transition and have described ritual assemblages or activities associated with early agricultural occupations (Gregory 2001; Miller in press). Operationalizing this idea is a challenge, but it bears further investigation. For example, in the southern Southwest, a number of artifacts that appear to have been used in ritual activities have been identified, including pipes and possibly cruciforms. Kisselberg (1993) and Heidke (2005) have argued that incipient plainwares represent small bowls that were likely used for rituals instead of as utilitarian vessels. Further studies of the context of these finds to determine if similarities exist across broad areas and techniques, such as use wear and residue analysis, may clarify the role of these items in ritual activities at early farming sites. This would be especially useful in examining the nature of *cerro de trincheras* sites, for if they served as ritual gathering spots versus refuges from conflict, then this should be reflected in the artifact assemblages.

The Role of Native Cultigens, Squash, and Beans in the Agricultural Transition

Current data indicate that squash and beans were later arrivals to the Southwest, with squash adopted around 1000 BC in many of the same settings where early maize was first used and beans arriving around 400 BC. The factors surrounding the adoption of these other cultigens should be investigated, especially in light of the fact that squash was domesticated in Mexico well before the time that maize first entered the Southwest (Smith 1997). As noted earlier, this may be the result of preservation issues, so continued analysis of a large number of flotation samples is needed to establish if this time discrepancy is valid. Experimental studies of growing techniques for squash and beans may establish the parameters necessary for their successful growth and the constraints surrounding their use. Studies of beans from upland rock shelters should be done to determine why they were used there and did not spread further until later in time, if, as current data indicate, beans were not adopted elsewhere until ceramic technology made their use more tenable. Finally, detailed botanical studies are needed to determine if wild plants like chenopodium were cultivated and/or tended. This may be reflected in both the genetic makeup of these plants or in other traits such as small changes in seed size. It is highly likely that wild-plant manipulation preceded the earliest use of maize, an idea supported by the recovery of cheno-ams associated with the earliest maize, but this should be established with direct evidence.

Agricultural Beginnings in the American Southwest

Agricultural beginnings in the American Southwest were not particularly glamorous. No roving hordes of farmers brought agriculture to the Southwest to save starving hunter-gatherers. Instead, some hunter-gatherers in well-watered settings with either kin or trade access to groups in Mexico who were farming maize learned about this new subsistence strategy during their interactions (Merrill et al. 2009). Some decided to adopt maize, probably because it fit well with their existing settlement and subsistence regime and because the environment allowed it. Others did not, and their story is equally interesting. Some groups, as is typical of hunter-gatherers cross-culturally, decided to invest more intensively in this subsistence strategy. These groups—evident primarily along the major waterways of southern Arizona—witnessed some major changes in adaptation tied to this decision to invest in agriculture, manifest archaeologically in the presence of irrigation systems and the built environment. Others, like those in the northern Southwest, adopted agriculture to suit their needs—either investing in storage technology to facilitate overwintering or intensifying the occupation of alluvial settings. The trajectory of change during the later (terminal) Late Archaic (Cienega phase in the south and Basketmaker in the north) was set by the decisions made by these early farmer-foragers. The environment and its constraints and possibilities clearly played a major role in this transition. But decision making by these hunter-gatherers and early farmers also played a crucial role in the cultural changes set forth by the initial decision to farm or not to farm.

The lesson from all of this is that agricultural beginnings cannot be explained with one answer or with many answers. It is context dependent and can only be understood within the physical and social environment in which it occurred, the history of previous developments within the region, and finally, the sometimes stochastic events that occurred that resulted in the knowledge that cultigens and cultivation techniques were available. The whys and hows of this subsistence decision appear to have been as variable as the subsequent impacts.

References

Adams, Jenny L.

1999 Refocusing the Role of Food-Grinding Tools as Correlates for Subsistence Strategies in the U.S. Southwest. *American Antiquity* 64:475–98.

2005 Early Agricultural Period Grinding Technology. In *Material Cultures and Lifeways of Early Agricultural Communities in Southern Arizona*, edited by R. Jane Sliva, pp. 99–119. Anthropological Papers No. 35. Center for Desert Archaeology, Tucson, AZ.

2015 San Pedro Phase Grinding Technology at Las Capas, AZ AA:12:111 (ASM). In *Implements of Change: Tools, Subsistence, and the Built Environment of Las Capas, an Early Agricultural Irrigation Community in Southern Arizona*, edited by James M. Vint, pp. 95–159. Anthropological Papers No. 51, Archaeology Southwest, Tucson.

Adams, Karen R.

1994 A Regional Synthesis of *Zea Mays* in the Prehistoric American Southwest. In *Corn and Culture in the Prehistoric New World*, edited by S. Johanneson and C. A. Hastorf, pp. 273–302. Westview Press, Boulder, CO.

Agenbroad, Larry D.

1966 Preliminary Report on a Desert Culture Site, San Pedro Valley, Arizona. Manuscript on file, Arizona State Museum Library, Tucson, AZ.

1970 Cultural Implications from the Statistical Analysis of a Prehistoric Lithic Site in Arizona. Unpublished M.A. thesis, Department of Anthropology, University of Arizona.

Agogino, George, and J. Hester

1956 A Re-evaluation of the San Jose Non-Ceramic Cultures. *El Palacio* 63:6–21.

Ames, Kenneth M.

1999 Myth of the Hunter Gatherer. *Archaeology* 52(5):44–49.

Amsden, Charles Avery

1935 The Pinto Basin Artifacts. In *The Pinto Basin Site*, by Elizabeth W. Crozer Campbell and William H. Campbell, pp. 33–51. Southwest Museum Papers No. 9, Los Angeles.

Amundsen-Meyer, Nicole Engel, and Sean Pickering (editors)
2011 *Identity Crisis: Archaeological Perspectives on Social Identity*. Proceedings of the 42nd Annual Chacmool Archaeology Conference, University of Calgary, Calgary, Alberta.

Antevs, Ernst
1948 Climatic Changes and Pre-White Man in the Great Basin, with Emphasis on Glacial and Postglacial Times. *University of Utah Bulletin* 38(20):168–91.
1955 Geologic-Climatic Dating in the West. *American Antiquity* 20:317–35.

Anthony, David W.
1990 Migration in Archeology: The Baby and the Bathwater. *American Anthropologist* 92:895–914.

Apicella, Coren L., Frank W. Marlowe, James H. Fowler, and Nicholas A. Christakis
2012 Social Networks and Cooperation in Hunter-Gatherers. *Nature* 482(7382):497–501.

Barceló, Juan A., Florencia Del Castillo Bernal, Ricardo del Olmo, Laura Mameli, F. J. Miguel Quesada, David Poza, and Xavier Vila
2014 Social Interaction in Hunter-gatherer Societies: Simulating the Consequences of Cooperation and Social Aggregation. *Social Science Computer Review* 32:417–36.

Bargman, Byrd A., Peggy A. Gerow, and Janette M. Elyea
1999 LA 110946, San Luis de Cabezon Site. In *Data Recovery along the 1995 Four Corners Pipeline; Sites in the San Juan Basin/Colorado Plateau, Sandoval, San Juan, McKinley Counties, New Mexico, Volume 2,* edited by Kenneth L. Brown, pp. 11–46. OCA/UNM Report No. 185–547D. Office of Contract Archaeology, University of New Mexico, Albuquerque.

Barker, Graeme
2009 *The Agricultural Revolution in Prehistory: Why Did Foragers Become Farmers?* Oxford University Press, Oxford, UK.

Barlow, K. Renee
2002 Predicting Maize Agriculture among the Fremont: An Economic Comparison of Farming and Foraging in the American Southwest. *American Antiquity* 67:65–88.

Basgall, Mark E., and M. C. Hall
2000 Morphological and Temporal Variation in Bifurcate-Stemmed Dart Points of the Western Great Basin. *Journal of California and Great Basin Anthropology* 22:237–76.

Bayham, Frank E., and Donald H. Morris
1986 Episodic Use of a Marginal Environment: A Synthesis. In *Prehistoric Hunter-Gatherers of South-Central Arizona: The Picacho Reservoir Archaic Project*, edited by Frank E. Bayham, Donald H. Morris, and M. Steven Shackley, pp. 359–81. Anthropological Field Studies No. 13. Arizona State University, Tempe.

Bayham, Frank E., Donald H. Morris, and M. Steven Shackley (editors)
1986 *Prehistoric Hunter-Gatherers of South-Central Arizona: The Picacho Reservoir Archaic Project*. Anthropological Field Studies No. 13. Arizona State University, Tempe.

Beale, Nicholas H.
2014 Social Interaction among Late Archaic and Incipient Agricultural Groups in the North American Southwest. Unpublished Ph.D. dissertation, Department of Anthropology, University of Oklahoma, Norman, OK.

Bean, Lowell J., and Katherine S. Saubel
1972 *Temalpakh: Cahuilla Indian Knowledge and Usage of Plants*. Malki Museum Press, Banning, California.

Beckes, M. R., and James M. Adovasio

1982 Perspectives on the Origins and Affinities of Mogollon on the Eastern Periphery. In *Mogollon Archaeology: Proceedings of the 1980 Mogollon Conference*, edited by Patrick H. Beckett, pp. 201–9. Acoma Books, Ramona, CA.

Bell, Willis H., and Edward F. Castetter

1937 *The Utilization of Mesquite and Screwbean by the Aborigines in the American Southwest.* Ethnobiological Studies in the American Southwest, Vol. V. The University of New Mexico Bulletin 5(2). University of New Mexico Press, Albuquerque, New Mexico.

Bernardini, Wesley

2005 Reconsidering Spatial and Temporal Aspects of Prehistoric Cultural Identity: A Case Study from the American Southwest. *American Antiquity* 70:31–54.

Berry, Claudia F., and Michael S. Berry

1986 Chronological and Conceptual Models of the Southwestern Archaic. In *Anthropology of the Desert West: Essays in Honor of Jesse D. Jennings*, edited by Carol J. Condie and Don D. Fowler, pp. 253–327. Anthropological Papers No. 110. University of Utah Press, Salt Lake City.

Berry, Michael S.

1982 *Time, Space and Transition in Anasazi Prehistory.* University of Utah Press, Salt Lake City.

Bettinger, Robert L.

2009 *Hunter-Gatherer Foraging: Five Simple Models.* Eliot Werner Publications, Clinton Corners, NY.

Binford, Lewis R.

1980 Willow Smoke and Dogs' Tails: Hunter-Gatherer Settlement Systems and Archaeological Site Formation. *American Antiquity* 45:4–20.

2001 *Constructing Frames of Reference, An Analytical Method for Archaeological Theory Building Using Ethnographic and Environmental Data Sets.* University of California Press, Berkeley.

Bird, Douglas W. and James F. O'Connell

2006 Behavioral Ecology and Archaeology. *Journal of Archaeological Research* 14:143–88.

Bird, Rebecca Bliege, Douglas W. Bird, Brian F. Codding, Christopher H. Parker, and James H. Jones

2008 The "Fire Stick Farming" Hypothesis: Australian Aboriginal Foraging Strategies, Biodiversity and Anthropogenic Fire Mosaics. *Proceedings of the National Academy of Sciences* 105(39):14796–801.

Bowles, Samuel

2011 Cultivation of Cereals by the First Farmers Was Not More Productive Than Foraging. *Proceedings of the National Academy of Sciences* 108(12):4760–65.

Browman, David L., Gayle J. Fritz, and Patty Jo Watson

2005 Origins of Food-Producing Economies in the Americas. In *The Human Past*, edited by Christopher Scarre, pp. 306–49. Thames and Hudson, London.

Brown, David E.

1982 Biotic Communities of the American Southwest—United States and Mexico. *Desert Plants* 4:1–4.

Bryan, Kirk, and J. H. Toulouse Jr.

1943 The San Jose Non-ceramic Culture and Its Relation to Puebloan Culture in New Mexico. *American Antiquity* 9:269–80.

Byrd, Rachael M.

2014 Phenotypic Variation of Transitional Forager-Farmers in the Sonoran Desert. *American Journal of Physical Anthropology* 155:579–90.

Cabana, Graciela S., and Jeffrey J. Clark (editors)

2011 *Rethinking Anthropological Perspectives on Migration.* University Press of Florida, Gainesville.

Cameron, Catherine M.

1995 Migration and the Movement of Southwestern Peoples. *Journal of Anthropological Archaeology* 14:104–24.

Campbell, Elizabeth W. Crozer, and William H. Campbell

1935 *The Pinto Basin Site.* Southwest Museum Papers No. 9, Los Angeles.

Carpenter, John, Guadalupe Sanchez, and Elisa Villalpando

1999 Preliminary Investigations at La Playa, Sonora, Mexico. *Archaeology Southwest* 13(1):10–11.

2002 Of Maize and Migration: Mode and Tempo in the Diffusion of *Zea Mays* in Northwest Mexico and the American Southwest. In *Traditions, Transitions, and Technologies, Themes in Southwestern Archaeology*, edited by Sarah H. Schlanger, pp. 245–58. University Press of Colorado, Boulder.

2005 The Late Archaic/Early Agricultural Period in Sonora, Mexico. In *New Perspectives on the Late Archaic across the Borderlands,* edited by Bradley J. Vierra, pp. 3–40. University of Texas Press, Austin.

In press The Archaic Period in Sonora. In *The Southwest Archaic*, edited by Bradley J. Vierra. University of Utah Press.

Cashdan, Elizabeth

1990 *Risk and Uncertainty in Tribal and Peasant Economies.* Westview Press, Boulder, CO.

Castetter, Edward F., and Willis H. Bell

1937 *The Aboriginal Utilization of the Tall Cacti in the American Southwest.* Bulletin 307, Biological Series 5(1). University of New Mexico Press, Albuquerque.

Castetter, Edward F., and Ruth M. Underhill

1935 *The Ethnobiology of the Papago Indians.* Bulletin 275, Biological Series 4(3). University of New Mexico Press, Albuquerque.

Charles, Mona C., and Sally J. Cole

2006 Chronology and Cultural Variation in Basketmaker II. *Kiva* 72:167–216.

Charles, Mona C., Leslie M. Sesler, and Timothy D. Hovezak

2006 Understanding Eastern Basketmaker II Chronology and Migrations. *Kiva* 72: 217–38.

Chavarria, Sara P.

1996 *Archaeological Investigations at the Summit at Alvernon Site, AZ BB:9:280 (ASM).* Report No. 4, Old Pueblo Archaeology Center Tucson, AZ.

Clark, Jeffrey J.

2001 *Tracking Prehistoric Migrations: Pueblo Settlers among the Tonto Basin Hohokam.* Anthropological Papers No. 65. University of Arizona Press, Tucson.

Coltrain, Joan B., Joel C. Janetski, and Shawn W. Carlyle

2007 Stable and Radio-Isotope Chemistry of Western Basketmaker Burials: Implications for Early Puebloan Diets and Origins. *American Antiquity* 72:301–21.

Cordell, Linda S. and Maxine E. McBrinn

2012 *Archaeology of the Southwest, 3rd Ed.* Left Coast Press, Walnut Creek, CA

Cummings, Linda Scott, Jammi L. Ladwig, and Chad Yost

2013 Phytolith Analysis of Maize Cob Fragments from Las Capas (AZ AA: 12:111), Tucson, Arizona. *The Digital Archaeological Record*, doi: 10.6067/XCV83B612B.

Czaplicki, Jon S.

1984 *A Class III Survey of the Tucson Aqueduct Phase A Corridor, Central Arizona Project: An Intensive Survey in the Lower Santa Cruz River Basin, Picacho Reservoir to Rillito, Arizona*. Archaeological Series No. 165. Arizona State Museum, Tucson.

Damp, Jonathan E.

2007 Zuni Emergent Agriculture: Economic Strategies and the Origins of Zuni. In *Zuni Origins: Toward a New Synthesis of Southwestern Archaeology*, edited by David A. Gregory and David R. Wilcox, pp. 118–32. University of Arizona Press, Tucson, AZ.

Damp, Jonathan E., Stephen A. Hall, and Susan J. Smith

2002 Early Irrigation on the Colorado Plateau near Zuni Pueblo, New Mexico. *American Antiquity* 67:665–76.

Dart, Allen

1986 *Archaeological Investigations at La Paloma: Archaic and Hohokam Occupations at Three Sites in the Northeastern Tucson Basin, Arizona*. Anthropological Papers No. 4. Institute for American Research, Tucson, AZ.

Dean, Rebecca M.

2005 Site-Use Intensity, Cultural Modification of the Environment, and the Development of Agricultural Communities in Southern Arizona. *American Antiquity* 70:403–31.

Del Bene, Terry A., and William R. Branchard

1994 A Synthesis of Archaic Archaeology on the Navajo Indian Irrigation Project, Northwestern New Mexico. In *Archaic Hunter-Gatherer Archaeology in the American Southwest*, edited by Bradley J. Vierra, pp. 231–90. Contributions in Anthropology 13(1), Eastern New Mexico University, Portales.

Del Castillo, Florencia, Juan Antonio Barcelo, Laura Mameli, F. J. Miguel Quesada, and Xavier Vila

2014 Modeling Mechanisms of Cultural Diversity and Ethnicity in Hunter-Gatherers. *Journal of Archaeological Method and Theory* 21:364–84.

Dello-Russo, Robert

2006 Got Water? Modeling the Early Use of Maize in Regions of the Southwest Gila Mountains and Beyond. *New Mexico Botanist: Natural History of the Gila—A Symposium*: 17–33.

Díaz-Andreu, Sam Lucy, Staša Babić, and David N. Edwards

2005 *The Archaeology of Identity*. Taylor and Francis, London; New York.

Dick, Herbert W.

1965 *Bat Cave*. School of American Research, Santa Fe, NM.

Diehl, Michael W.

1992 Architecture as a Material Correlate of Mobility Strategies: Some Implications for Archaeological Interpretation. *Cross-Cultural Research* 26:1–35.

1997 Rational Behavior, the Adoption of Agriculture, and the Organization of Subsistence during the Late Archaic Period in the Greater Tucson Basin. In *Rediscovering Darwin: Evolutionary Theory and Archaeological Explanation*, edited by C. M. Barton and G. A. Clark, pp. 251–65. Archaeological Papers No. 7. American Anthropological Association, Arlington, VA.

2005a Morphological Observations on Recently Recovered Early Agricultural Period Maize Cob Fragments from Southern Arizona. *American Antiquity* 70:361–75.

2005b When Corn Was Not Yet King. In *Subsistence and Resource Use Strategies of Early Agricultural Communities in Southern Arizona*, edited by Michael W. Diehl, pp. 1–18. Anthropological Papers No. 34. Center for Desert Archaeology, Tucson, AZ.

2005c Epilogue: "Farmaging" during the Early Agricultural Period. In *Subsistence and Resource Use Strategies of Early Agricultural Communities in Southern Arizona*, edited by Michael W. Diehl, pp. 181–84. Anthropological Papers No. 34. Center for Desert Archaeology, Tucson, AZ.

2015 Foraging off the Gardens: Diet Breadth and Anthropogenic Landscape Change during the Early Agricultural Period in the Tucson Basin. In *Implements of Change: Tools, Subsistence, and the Built Environment of Las Capas, an Early Agricultural Irrigation Community in Southern Arizona*, edited by James M. Vint, pp. 345–75. Anthropological Papers No. 51, Archaeology Southwest, Tucson.

Diehl, Michael W. and David A. Gregory

2002 Duration, Continuity, and Intensity of Occupation at a Late Cienega Phase Settlement in the Santa Cruz River Floodplain. In *Traditions, Transitions, and Technologies: Themes in Southwestern Archaeology*, edited by Sarah H. Schlanger, pp. 200–223. University Press of Colorado, Boulder.

Diehl, Michael W., and Jennifer A. Waters

2006 Aspects of Optimization and Risk during the Early Agricultural Period in Southeastern Arizona. In *Behavioral Archaeology and the Transition to Agriculture*, edited by Douglas Kennett and Bruce Winterhalder, pp. 63–86. University of California Press, Berkeley.

Doleman, William H.

2005 Environmental Constraints on Forager Mobility and the Use of Cultigens in Southeastern Arizona and Southern New Mexico. In *The Late Archaic across the Borderlands: From Foraging to Farming*, edited by Bradley J. Vierra, pp. 113–40. University of Texas Press, Austin.

Douglas, John E., and Douglas B. Craig

1986 *Investigations of Archaic and Hohokam Sites on the Flying V Ranch, Tucson, Arizona*. Archaeological Report No. 13. Pima Community College, Tucson, AZ.

Downum, Christian E., Adrianne G. Rankin, and Jon S. Czaplicki

1986 *A Class III Archaeological Survey of the Phase B Corridor, Tucson Aqueduct, Central Arizona Project*. Archaeological Series No. 168. Arizona State Museum, Tucson.

Duff, Andrew I.

1998 The Process of Migration in the Late Prehistoric Southwest. In *Migration and Reorganization: The Pueblo IV Period in the American Southwest*, edited by K. Spielmann, pp. 31–52. Anthropological Research Papers No. 51. Arizona State University, Tempe.

Dunbier, Roger

1968 *The Sonoran Desert: Its Geography, Economy and People*. University of Arizona Press, Tucson.

Eddy, Frank W.

1958 A Sequence of Cultural and Alluvial Deposits in the Cienega Creek Basin, Southeastern Arizona. Unpublished M.A. Thesis, Department of Anthropology, University of Arizona, Tucson.

1961 *Excavations at Los Pinos Phase Sites in the Navajo Reservoir District.* Museum of New Mexico Papers in Anthropology No. 4. Museum of New Mexico Press, Santa Fe.

1966 *Prehistory in the Navajo Reservoir District, Northwestern New Mexico.* Museum of New Mexico Papers in Anthropology No. 15(1–2). Museum of New Mexico Press, Santa Fe.

Eddy, Frank W., and M. E. Cooley

1983 *Cultural and Environmental History of Cienega Valley, Southeastern Arizona.* University of Arizona Press, Tucson.

Elston, Robert G., and David W. Zeanah

2002 Thinking outside the Box: A New Perspective on Diet Breadth and Sexual Division of Labor in the Prearchaic Great Basin. *World Archaeology* 34:103–30.

Elyea, Janette M., and Patrick Hogan

1993 *Data Recovery at Three Archaic Sites on the Bolack Land Exchange.* University of New Mexico Office of Contract Archaeology, Albuquerque.

Ezzo, Joseph A., and William L. Deaver

1998 *Watering the Desert: Late Archaic Farming at the Costello-King Site.* Technical Series No. 68. Statistical Research, Tucson, AZ.

Ezzo, Joseph A., and Mary C. Stiner

2000 A Late Archaic Period Dog Burial from the Tucson Basin, Arizona. *Kiva* 66:291–305.

Fish, Paul R., Suzanne K. Fish, Austin Long, and Charles Miksicek

1986 Early Corn Remains from Tumamoc Hill, Southern Arizona. *American Antiquity* 51:563–72.

Fish, Suzanne K., Paul R. Fish, Gary Christopherson, Todd A. Pitezel, and James T. Watson

2011 Two Villages on Tumamoc Hill. *Journal of Arizona Archaeology* 1:185–96.

Fitzhugh, Ben

2001 Risk and Invention in Human Technological Evolution. *Journal of Anthropological Archaeology* 20:125–67.

Fonseca, Rute R. da, Bruce D. Smith, Nathan Wales, Enrico Cappellini, Pontus Skoglund, Matteo Fumagalli, José Alfredo Samaniego, Christian Carøe, María C. Ávila-Arcos, David E. Hufnagel, Thorfinn Sand Korneliussen, Filipe Garrett Vieira, Mattias Jakobsson, Bernardo Arriaza, Eske Willerslev, Rasmus Nielsen, Matthew B. Hufford, Anders Albrechtsen, Jeffrey Ross-Ibarra, M. Thomas, P. Gilbert

2015 The Origin and Evolution of Maize in the Southwestern United States. *Nature Plants* 1(1):1–5.

Ford, Richard I.

1981 Gardening and Farming before A.D. 1000: Patterns of Prehistoric Cultivation North of Mexico. *Journal of Ethnobiology* 1:6–27.

2013 The Cultural Ecology of Jemez Cave. In *From Mountaintop to Valley Bottom: Understanding Past Land-Use in the Northern Rio Grande, New Mexico*, edited by Bradley J. Vierra, pp. 69–79. University of Utah Press, Salt Lake City.

Formby, D. E.

1986 Pinto-Gypsum Complex Projectile Points from Arizona and New Mexico. *Kiva* 51:99–127.

Fowler, Catherine S.

1992 *In the Shadow of Fox Peak: An Ethnography of the Cattail-Eater Northern Paiute People of Stillwater Marsh.* Cultural Resource Series No. 5. US Fish and Wildlife Service, Stillwater National Wildlife Refuge, Fallon, NV.

1996 Historical Perspectives on Timbisha Shoshone Land Management Practices, Death Valley, California. In *Case Studies in Environmental Archaeology*, edited by E. J. Reitz, L. A. Newsom, and S. J. Scudd, pp. 87–102. Plenum Press, New York.

Freeman, Andrea K.

1997a *Archaeological Investigations at the Wetlands Site, AZ AA:12:90 (ASM).* Technical Report 97-5. Center for Desert Archaeology, Tucson, AZ.

1997b *Middle to Late Holocene Stream Dynamics of the Santa Cruz River, Tucson, Arizona: Implications for Human Settlement, the Transition to Agriculture, and Archaeological Site Preservation.* Unpublished Ph.D. dissertation, Department of Anthropology, University of Arizona, Tucson.

Freeman, Jacob

2012 Alternative Adaptive Regimes for Integrating Foraging and Farming Activities. *Journal of Archaeological Science* 39:3008–17.

Geib, Phil R.

1996 *Glen Canyon Revisited.* Anthropological Papers No. 119. University of Utah Press, Salt Lake City.

2002 Basketmaker II Horn Flaking Tools and Dart Point Production: Technological Change at the Agricultural Transition. In *Traditions, Transitions, and Technologies*, edited by Sarah H. Schlanger, pp. 272–306. University Press of Colorado, Boulder, CO.

2011 *Foragers and Farmers of the Northern Kayenta Region. Excavations along the Navajo Mountain Road.* University of Utah Press, Salt Lake City.

Geib, Phil R., and Edward Jolie

2008 The role of Basketry in Early Holocene Small Seed Exploitation: Implications of a Ca. 9,000 Year-Old Basket from Cowboy Cave, Utah. *American Antiquity* 73:83–102.

Geib, Phil R., and Kimberly Spurr

2000 The Forager to Farmer Transition on the Rainbow Plateau. In *Foundations of Anasazi Culture: The Basketmaker-Pueblo Transition,* edited by Paul F. Reed, pp. 175–200. University of Utah Press, Salt Lake City.

Gilman, Patricia A.

1997 *Wandering Villagers: Pit Structures, Mobility and Agriculture in Southeastern Arizona.* Anthropological Research Papers No. 49. Arizona State University, Tempe.

Gilman, Patricia A., Elizabeth M. Toney, and Nicholas H. Beale

2013 An Historical Ecological Perspective on Early Agriculture in the North American Southwest and Northwest Mexico. In *The Archaeology and Historical Ecology of Small Scale Economies*, edited by Victor D. Thompson and James C. Waggoner. University Press of Florida, Gainesville.

Gilpin, Dennis

1994 Lukachukai and Salina Springs: Late Archaic/Early Basketmaker Habitation Sites in the Chinle Valley, Northeastern Arizona. *Kiva* 60:203–18.

Graves, W. M., R. M. Wegener, and R. Ciolek-Torrello

2011 Settlement and Production at the Cashion Site Complex: The Early Archaic and Hohokam Periods at AZ T:11:94 (ASM). *Journal of Arizona Archaeology* 1(2):148–61.

Graves, Tim, Myles Miller, and Melinda Landreth

2014 *Three Millennia of Village Settlements in the Western Tularosa Basin: Excavations at Four Sites near the Organ Mountains on Fort Bliss in Doña Ana County, New Mexico.* QRI/GMI Joint Venture Report No. 25-005. Fort Bliss, El Paso, TX.

Greaves, Russell D., and Karen L. Kramer

2014 Hunter-Gatherer Use of Wild Plants and Domesticates: Archaeological Implications for Mixed Economies before Agricultural Intensification. *Journal of Archaeological Science* 41:263–71.

Gregory, David A. (editor)

1999 *Excavations in the Santa Cruz River Floodplain: The Middle Archaic Component at Los Pozos.* Anthropological Papers No. 20. Center for Desert Archaeology, Tucson.

2001 *Excavations in the Santa Cruz River Floodplain: The Early Agricultural Occupation at Los Pozos (AZ AA:12:91, ASM).* Archaeological Papers No. 21. Center for Desert Archaeology, Tucson, AZ.

Gregory, David A., and Bruce B. Huckell

1998 Small Round and Stable: Early Agricultural Period Pithouse Architecture and its Implications for the Rise of Long-Term Settlements in Southeastern Arizona. Paper presented at the 63rd Annual Meeting of the Society for American Archaeology, Seattle.

Gregory, David A., and Fred L. Nials

2005 The Environmental Context of Early Agricultural Period Occupations in the Tucson Basin. In *Subsistence and Resource Use Strategies of Early Agricultural Communities in Southern Arizona,* edited by Michael W. Diehl, pp. 19–71. Anthropological Papers No. 34. Center for Desert Archaeology, Tucson, AZ.

Gregory, David A., Michelle N. Stevens, Fred L. Nials, Mark R. Schurr, and Michael W. Diehl

2007 *Excavations in the Santa Cruz River Floodplain: Further Investigations at Los Pozos.* Anthropological Papers No. 34. Center for Desert Archaeology, Tucson, AZ.

Gremillion, Kristen J., and Dolores R. Piperno

2009 Human Behavioral Ecology, Phenotypic (Developmental) Plasticity, and Agricultural Origins. *Current Anthropology* 50:615–19.

Guernsey, Samuel J., and Alfred V. Kidder

1921 *Basket-Maker Caves of Northeastern Arizona: Report on the Explorations, 1916–17.* Peabody Museum of American Archaeology and Ethnography, Harvard University, Cambridge, MA.

Gurven, Michael, and Kim Hill

2009 Why Do Men Hunt? A Reevaluation of "Man the Hunter" and the Sexual Division of Labor. *Current Anthropology* 50:51–74.

Hackbarth, Mark R.

1998 *Archaic and Hohokam Occupation of the Mayo Boulevard Project Area in Northeast Phoenix, Arizona.* Anthropological Papers No. 8. Pueblo Grande Museum, Phoenix, AZ.

2001 Late Archaic and Red Mountain Phase Task Group Size in Paradise Valley, Arizona: Evidence from the Last Ditch Site. *Kiva* 67:81–106.

Halbirt, Carl D., and James Copus

1993 Tator Hills (AZ AA:6:18[ASM]), A Multicomponent Campsite. In *Archaic Occupation on the Santa Cruz Flats: The Tator Hills Archaeological Project,* edited by Carl D. Halbirt and T. Kathleen Henderson, pp. 15–53. Report submitted to the U. S. Department of the Interior, Bureau of Reclamation, Arizona Projects office, Contract No. 3-PA-30-00740. Northland Research, Inc., Flagstaff, AZ.

Hall, Stephen A.

2010 Early Maize Pollen from Chaco Canyon, New Mexico, USA. *Palynology* 34:125–37.

Hall, John D.

In press The Phoenix Basin Archaic. In *The Southwest Archaic,* edited by Bradley J. Vierra. University of Utah Press.

Hall, John D., and Robert M. Wegener (editors)

2014 *Project Background and Excavation Results.* The Luke Air Force Base Solar-Power Array Archaeological Project: 5000 Years of Aboriginal Land Use in the Western Phoenix Basin, Vol. 1. Technical Report 14-18, Statistical Research, Inc., Tucson, AZ.

Halstead, Paul, and John O'Shea

1989 Introduction: Cultural Responses to Risk and Adversity. In *Bad Year Economics: Cultural Responses to Risk and Uncertainty,* edited by Paul Halstead and John O'Shea, pp. 1–7. Cambridge University Press, Cambridge, UK.

Hard, Robert J.

1990 Agricultural Dependence in the Mountain Mogollon. In *Perspectives on Southwestern Prehistory,* edited by Paul R. Minnis and Charles L. Redman, pp. 135–49. Westview Press, Boulder, CO.

Hard, Robert J., Raymond P. Mauldin, and Gerry R. Raymond

1996 Mano Size, Stable Carbon Isotope Ratios, and Macrobotanical Remains as Multiple Lines of Evidence of Maize Dependence in the American Southwest. *Journal of Archaeological Method and Theory* 3:253–318.

Hard, Robert J., and John R. Roney

1998 A Massive Terraced Village Complex in Chihuahua, Mexico, 3000 Years before Present. *Science* 279(5357):1661–64.

2004 Late Archaic Hilltop Settlements in Northwest Chihuahua, Mexico. In *Identity, Feasting, and the Archaeology of the Greater Southwest,* edited by Barbara J. Mills, pp. 276–94. University Press of Colorado, Boulder, CO.

2005 The Transition to Farming on the Rio Casa Grandes and in the Southern Jornada Mogollon Region. In *The Late Archaic across the Borderlands: From Foraging to Farming,* edited by Bradley J. Vierra, pp. 141–86. University of Texas Press, Austin.

2007 Cerros de Trincheras in Northwestern Chihuahua: The Arguments for Defense. In *Trincheras Sites in Time, Place, and Society,* edited by Suzanne K. Fish, Paul R. Fish, and M. Elisa Villalpando, pp. 11–52. University of Arizona Press, Tucson.

Harrington, Mark Raymond
1957 *A Pinto Site at Little Lake California.* Southwest Museum Papers No. 17, Los Angeles.
Haury, Emil W.
1957 An Alluvial Site on the San Carlos Indian Reservation, Arizona. *American Antiquity* 23:2–27.
1962 The Greater American Southwest. In *Courses Toward Urban Life: Some Archaeological Considerations of Cultural Alternatives*, edited by Robert J. Braidwood and Gordon R. Willey, pp. 106–31. Publications in Anthropology No. 32. The Viking Fund, Wenner-Gren Foundation, New York.
1976 *The Hohokam: Desert Farmers and Craftsmen.* University of Arizona Press, Tucson.
Hawkes, Kristen, and Rebecca Bliege Bird
2002 Showing Off, Handicap Signaling, and the Evolution of Men's Work. *Evolutionary Anthropology* 11:58–67.
Haynes, C. Vance Jr.
1968 Geochronology of Late-Quaternary Alluvium. In *Means of Correlation of Quaternary Successions, Vol. 8,* pp. 591–631. Proceedings of the VII Congress, International Association for Quaternary Research, University of Utah Press, Salt Lake City.
Heidke, James M.
1999 Cienega Phase Incipient Plain Ware from Southeastern Arizona. *Kiva* 64:311–38.
2005 Incipient Plain Ware Type Descriptions. In *Material Cultures and Lifeways of Early Agricultural Communities in Southern Arizona,* edited by R. Jane Sliva, pp. 249–55. Anthropological Papers No. 35. Center for Desert Archaeology, Tucson.
2006 Native American Pottery. In *Rio Nuevo Archaeology, 2000–2003: Investigations at the San Agustin Mission and Mission Gardens, Tucson Presidio, Tucson Pressed Brick Company, and Clearwater Site,* edited by J. H. Thiel and J. B. Mabry, pp. 7.1–7.93. Technical Report No. 2004-11. Desert Archaeology Inc., Tucson, AZ.
2015 San Pedro and Early Cienega Phase Clay Objects and Hohokam Pottery from Las Capas, AZ AA:12:111 (ASM): Dating, Technology, Use, and Discard. In *Implements of Change: Tools, Subsistence, and the Built Environment of Las Capas, an Early Agricultural Irrigation Community in Southern Arizona*, edited by James Vint, pp. 161–226. Anthropological Papers No. 51, Archaeology Southwest, Tucson.
Heizer, Robert F., and Alex D. Krieger
1956 *The Archaeology of Humboldt Cave, Churchill County, Nevada.* University of California Publications in American Archaeology and Ethnology 47(1), Berkeley.
Hendrickson, Dean A., and W. L. Minckley
1985 Cienegas, Vanishing Climax Communities of the American Southwest. *Desert Plants* 6:131–75.
Hesse, S. Jerome, India S. Hesse, and Stephanie M. Whittlesey
2010 Overview of Excavation Results and Site Structure. In *Recurrent Sedentism and the Making of Place: Archaeological Investigations at Las Capas, A Preceramic Farming Community in the Tucson Basin, Southern Arizona,* edited by Stephanie M. Whittlesey, S. Jerome Hesse, and Michael S. Foster. Report on file, Pima County Cultural Resources and Historic Preservation Office, Tucson, Arizona.
Hill, Jane H.
2001 Proto-Uto-Aztecan: A Community of Cultivators in Central Mexico? *American Anthropologist* 103:913–34.

2002 Toward a Linguistic Prehistory of the Southwest: "Azteco-Tanoan" and the Arrival of Maize Cultivation. *Journal of Anthropological Research* 58:457–75.

Hogan, Patrick

1994 Foragers to Farmers II: A Second Look at the Adoption of Agriculture in Northwestern New Mexico. In *Archaic Hunter-Gatherer Archaeology in the American Southwest*, edited by Bradley J. Vierra, pp. 155–84. Contributions in Anthropology 13(1), Eastern New Mexico University, Portales.

Holmer, Richard N.

1986 Common Projectile Points of the Intermountain West. In *Anthropology of the Desert West: Essays in Honor of Jesse D. Jennings*, edited by Carol J. Condie and Don D. Fowler, pp. 89–115. Anthropological Papers No. 110. University of Utah Press, Salt Lake City.

Huber, Edgar K.

2005 Early Maize at the Old Corn Site (LA 137258). In *Fence Lake Project: Archaeological Data Recovery in the New Mexico Transportation Corridor and First Five-Year Permit Area, Fence LakeCoal Mine Project, Catron County, New Mexico*, edited by Edgar K. Huber and Carla R. Van West, pp. 36.1–36.33. Technical Series 84, Statistical Research Inc., Tucson, AZ.

Huckell, Bruce B.

1984 *The Archaic Occupation of the Rosemont Area, Northern Santa Rita Mountains, Southeastern Arizona*. Archaeological Series 147(1). Arizona State Museum, Tucson.

1990 *Late Preceramic Farmer-foragers in Southeastern Arizona: A Cultural and Ecological Consideration of the Spread of Agriculture into the Arid Southwestern United States*. Unpublished Ph.D. Dissertation. Department of Arid Lands Resource Studies, University of Arizona, Tucson.

1995 *Of Marshes and Maize: Preceramic Agricultural Settlements in the Cienega Valley, Southeastern Arizona*. Anthropological Papers No. 59. University of Arizona Press, Tucson.

1996a The Archaic Prehistory of the North American Southwest. *Journal of World Prehistory* 10:305–73.

1996b Middle to Late Holocene Stream Behavior and the Transition to Agriculture in Southeastern Arizona. In *Early Formative Adaptations in the Southern Southwest*, edited by Barbara J. Roth, pp. 27–36. Monographs in World Archaeology No. 25. Prehistory Press, Madison, WI.

Huckell, Bruce B., and Lisa W. Huckell

1984 *Excavations at Milagro, a Late Archaic Site in the Eastern Tucson Basin*. Manuscript on file, Arizona State Museum, University of Arizona, Tucson.

Huckell, Bruce B., Lisa W. Huckell, and Karl K. Benedict

2002 Maize Agriculture and the Rise of Mixed Farming-Foraging Economies in Southeastern Arizona during the Second Millenium BC. In *Traditions, Transitions, and Technologies*, edited by Sarah H. Schlanger, pp. 137–59. University Press of Colorado, Boulder.

Huckell, Bruce B., Lisa W. Huckell, and Suzanne K. Fish

1995 *Investigations at Milagro, a Late Preceramic Site in the Eastern Tucson Basin*. Technical Report No. 94-5. Center for Desert Archaeology, Tucson, AZ.

Huckell, Bruce B., Lisa W. Huckell, and M. S. Shackley

1999 McEuen Cave. *Archaeology Southwest* 13(1):12.

Huckell, Lisa W.
2006 Ancient Maize in the American Southwest: What Does It Look Like and What Can It Tell Us? In *Histories of Maize: Multidisciplinary Approaches to the Prehistory, Biogeography, Domestication, and Evolution of Maize*, edited by J. E. Staller, R. H. Tykot, and B. F. Benz, pp. 97–107. Elsevier Academic Press, San Diego.

Hunter-Anderson, Rosalind
1986 *Prehistoric Adaptation in the American Southwest*. Cambridge University Press, New York.

Hyland, David C., James M. Adovasio, and J. S. Illingworth
2003 The Perishable Artifacts. In *Pendejo Cave*, edited by R. S. MacNeish and J. G. Libby, pp. 297–416. University of New Mexico, Albuquerque.

Irwin Williams, Cynthia
1967 Picosa: The Elementary Southwestern Culture. *American Antiquity* 32:441–57.
1968 *Early Man in Western North America*. Contributions in Anthropology 1(4). Eastern New Mexico University, Portales.
1973 *The Oshara Tradition: Origins of Anasazi Culture*. Contributions to Anthropology 5(1), Eastern New Mexico University, Portales.

Jennings, Jesse D.
1957 *Danger Cave*. Anthropological Papers No. 27. Department of Anthropology, University of Utah, Salt Lake City.

Johnson, Amber L.
2014 Exploring Adaptive Variation among Hunter-gatherers with Binford's Frames of Reference. *Journal of Archaeological Research* 22:1–42.

Jones, Jeffrey T., and Barbara K. Montgomery
2014 *Epley's Ruin: A Ballcourt Village in the Safford Valley*. Tierra Archaeological Report No. 2014-066, Tucson, AZ.

Justice, Noel D.
2002 *Stone Age Spear and Arrow Points of the Southwestern United States*. Indiana University Press, Bloomington.

Kaldahl, Eric J.
2001 *The Rosetta Site: Archaic Occupations in the Southern Santa Catalina Foothills*. Old Pueblo Archaeology Center, Tucson, AZ.

Kantner, John
2004 *Ancient Puebloan Southwest*. Cambridge University Press, Cambridge, UK.

Kaplan, Lawrence
1965 Archaeology and Domestication in American Phaseolus (Beans). *Economic Botany* 19:358–68.

Kearns, Timothy M.
In press Archaic Time and Distance in the San Juan Basin. In *The Southwest Archaic*, edited by Bradley J. Vierra. University of Utah Press.

Kearns, Timothy M., Chris A. Kugler, and Paul Stirniman
1998 The Dog Leg Site (LA 6448): A Basketmaker II Cache Locale in the Southern Chuska Valley, New Mexico. In *Pipeline Archaeology 1990–1993: The El Paso Natural Gas North System Expansion Project, New Mexico and Arizona: Volume 2: Archaic, Basketmaker II, Protohistoric, and Aceramic Sites in Northwest New Mexico*, edited by T. G. Baugh, T. M. Kearns and C. W. Wheeler. Western Cultural Resource Management Inc., Farmington, NM.

Kearns, Timothy M., and Janet L. McVickar (editors)
2007 *Time, Place, and Society: Project Synthesis.* Pipeline Archaeology 1990–1993: The El Paso Natural Gas North System Expansion Project, New Mexico and Arizona, Vol. XIII. Western Cultural Resource Management, Farmington, NM.

Keeley, Lawrence H.
1995 Protoagricultural Practices among Hunter-Gatherers, A Cross-Cultural Survey. In *Last Hunters, First Farmers: New Perspectives on the Prehistoric Transition to Agriculture*, edited by T. Douglas Price and Anne Birgitte Gebauer, pp. 243–99. School of American Research Press, Santa Fe, NM.

Kelly, Isabel T.
1964 *Southern Paiute Ethnography.* Anthropological Papers No. 69. University of Utah, Salt Lake City.

Kelly, Isabel T., and Catherine S. Fowler
1986 Southern Paiute. In *Handbook of North American Indians, Vol. 11: Great Basin.* Edited by W. L. d'Azevedo, pp. 368–79. Smithsonian Institution, Washington, DC.

Kelly, Robert L.
2013 *The Lifeways of Hunter-Gatherers: The Foraging Spectrum.* Cambridge University Press, New York.

Kelly, William H.
1977 *Cocopa Ethnography.* Anthropological Papers of the University of Arizona 29. University of Arizona Press, Tucson.

Kemp, Brian M., A. Gonzalez-Olivier, R. S. Mathis, C. Monroe, K. B. Schroeder, J. McDonough, G. Rhett, A. Resendez, R. I. Peñaloza-Espinosa, L. Buentello-Malo, C. Gorodesky, and D. G. Smith
2010 Evaluating the Farming/Language Dispersal Hypothesis with Genetic Variation Exhibited by Populations in the Southwest and Mesoamerica. *Proceedings of the National Academy of Sciences* 107(15):6759–64.

Kennett, Douglas J., Brendan J. Culleton, James P. Kennett, Jon M. Erlandson, and Kevin G. Cannariato
2007 Middle Holocene Climate Change and Human Population Dispersal in Western North America. In *Climate Change and Cultural Dynamics: A Global Perspective on Mid-Holocene Transitions*, edited by David G. Anderson, Kirk A. Maasch, and Daniel H. Sandweiss, pp. 531–57. Elsevier, New York.

Kennett, Douglas J., and Bruce Winterhalder (editors)
2006 *Behavioral Ecology and the Transition to Agriculture.* University of California Press, Oakland.

Kidder, Alfred V.
1927 Southwestern Archaeological Conference. *Science* 66:489–91.

Kidder, Alfred V., and Samuel J. Guernsey
1919 *Archaeological Explorations in Northeastern Arizona.* Bureau of American Ethnology Bulletin 65. Washington, DC.

Kisselberg, JoAnn E.
1993 The Tator Hills Ceramics: Perspectives on a Developing Technology. In *Archaic Occupation on the Santa Cruz Flats: The Tator Hills Archaeological Project*, edited by Carl D. Halbirt and T. Kathleen Henderson, pp. 279–304. Report submitted to U.S. Department of the Interior, Bureau of Reclamation, Arizona Projects Office. Northland Research Inc., Flagstaff, AZ.

Kohler, Tim A., Matt Glaude, Jean-Pierre Bocquet-Appel, and Brian M. Kemp

2008 The Neolithic Demographic Transition in the U.S. Southwest. *American Antiquity* 73:645–69.

Kuijt, Ian

1996 Negotiating Equality through Ritual: A Consideration of Late Natufian and Pre-Pottery Neolithic A Period Mortuary Practices. *Journal of Anthropological Archaeology* 15:313–36.

2000 *Life in Neolithic Farming Communities: Social Organization, Identity, and Differentiation.* Kluwer Academic/Plenum Publishers, New York.

Larson, Mary Lou

1997 Housepits and Mobile Hunter-Gatherers: A Consideration of the Wyoming Evidence. *Plains Anthropologist* 42:353–69.

Laumbach, Karl

n.d. The Cañada Alamosa Project. http://www.canadaalamosaproject.org

LeBlanc, Steven A.

2008 The Case for an Early Farmer Migration into the Greater American Southwest. In *Archaeology without Borders*, edited by Laurie D. Webster and Maxine E. McBrinn, pp. 107–42. University Press of Colorado, Boulder.

LeBlanc, Steven A., C. G. Turner, and M. E. Morgan

2007 Genetic Relationships Based on Discrete Dental Traits: Basketmaker II and Mimbres. *International Journal of Osteoarchaeology* 18(2):109–30.

Lentz, Stephen C.

2006 *High Rolls Cave: Archaic Subsistence in Southern New Mexico.* Office of Archaeological Studies, Museum of New Mexico, Santa Fe.

Liljeblad, Sven, and Catherine S. Fowler

1986 Owens Valley Paiute. In *Handbook of North American Indians, Vol. 11: Great Basin,* edited by W. L. d'Azevedo, pp. 412–34. Smithsonian Institution, Washington, DC.

Lipe, William D.

1999 Basketmaker II (1000 B.C.–A.D. 500). In *Colorado Prehistory: A Context for the Southern Colorado River Basin,* edited by William D. Lipe, Mark D. Varien, and Richard H. Wilshusen, pp. 132–65. Colorado Council of Professional Archaeologists, Denver.

Louderback, Lisbeth A., Donald K. Grayson, and Marcos Llobera

2011 Middle Holocene Climates and Human Population Densities in the Great Basin, Western USA. *The Holocene* 21:366–73.

Mabry, Jonathan B.

1990 *A Late Archaic Occupation at AZ AA:12:105 (ASM).*Technical Report No. 90-6. Center for Desert Archaeology, Tucson, AZ.

1996 *Canals and Communities: Small-Scale Irrigation Systems.* University of Arizona Press, Tucson.

1997 Rewriting Prehistory: Recent Discoveries at Cienega Phase Sites in the Santa Cruz Floodplain. *Archaeology in Tucson: Newsletter of the Center for Desert Archaeology* 11(3):1.

1998a *Archaeological Investigations of Early Village Sites in the Middle Santa Cruz Valley: Analysis and Synthesis.* Anthropological Paper No. 19. Center for Desert Archaeology, Tucson, AZ.

1998b *Paleoindian and Archaic Sites in Arizona.* Report prepared for the Arizona State Historical Preservation Office, Phoenix, AZ.

2002 The Role of Irrigation in the Transition to Agriculture and Sedentism in the Southwest: A Risk Management Model. In *Traditions, Transitions, and Technologies: Themes in Southwestern Archaeology*, edited by Sarah H. Schlanger, pp. 178–99. University Press of Colorado, Boulder.

2005a Changing Knowledge and Ideas about the First Farmers in Southeastern Arizona. In *The Late Archaic across the Borderlands*, edited by Bradley J. Vierra, pp. 41–83. University of Texas Press, Austin.

2005b Diversity in Early Southwestern Farming and Optimization Models of Transitions to Agriculture. In *Subsistence and Resource Use Strategies of Early Agricultural Communities in Southern Arizona*, edited by Michael W. Diehl, pp. 113–52. Anthropological Papers No. 34. Center for Desert Archaeology, Tucson, AZ.

2006 Radiocarbon Dating of the Early Occupations. In *Rio Nuevo Archaeology, 2000–2003: Investigations at the San Agustin Mission and Mission Gardens, Tucson Presidio, Tucson Pressed Brick Company, and Clearwater Site*, edited by J. Homer Thiel and Jonathan B. Mabry, pp. 19.1–19.5. Technical Report No. 2004-11. Desert Archaeology, Inc., Tucson, AZ.

2008 *Las Capas: Early Irrigation and Sedentism in a Southwestern Floodplain*. Anthropological Papers No. 28. Center for Desert Archaeology, Tucson, AZ.

Mabry, Jonathan B., John P. Carpenter, and Guadalupe Sanchez

2008 Archaeological Models of Early Uto-Aztecan Prehistory in the Arizona-Sonora Borderlands. In *Archaeology without Borders: Contact, Commerce, and Change in the U. S. Southwest and Northwestern Mexico*, edited by Laurie D. Webster and Maxine E. McBrinn, pp. 155–84. University Press of Colorado, Boulder.

Mabry, Jonathan B. and William E. Doolittle

2008 Modeling the Early Agricultural Frontier in the Desert Borderlands. In *Archaeology without Borders: Contact, Commerce, and Change in the U. S. Southwest and Northwestern Mexico*, edited by Laurie D. Webster and Maxine E. McBrinn, pp. 55–70. University Press of Colorado, Boulder.

MacNeish, Richard S.

1958 Preliminary Archaeological Investigations in the Sierra de Tamaulipas, Mexico. *Transaction of the American Philosophical Society* 48:6.

1993 *Preliminary Investigations of the Archaic in the Region of Las Cruces, New Mexico*. U.S. Army Air Defense Artillery Center, Fort Bliss, TX.

MacNeish, Richard S., and Patrick H. Beckett

1987 *The Archaic Chihuahua Tradition*. COAS Publishing, Las Cruces, NM.

Madsen, David B., and Dave N. Schmitt

1998 Mass Collecting and the Diet Breadth Model: A Great Basin Example. *Journal of Archaeological Science* 25:445–55.

Malhi, R. S., H. M. Mortensen, J. A. Eshleman, J. G. Lorenz, F. A. Kaestle, J. R. Johnson, C. Gorodesky, and D. G. Smith

2003 Native American mtDNA prehistory in the American Southwest. *American Journal of Physical Anthropology* 120:108–24.

Marlowe, Frank W.

2007 Hunting and Gathering: The Human Sexual Division of Foraging Labor. *Cross-Cultural Research* 41:170–95.

Martin, Paul S., Elaine Bluhm, Hugh C. Cutler, Roger T. Grange, and John B. Rinaldo

1952 *Mogollon Cultural Continuity and Change; The Stratigraphic Analysis of Tularosa and Cordova Caves*. Fieldiana: Anthropology Volume 40. Chicago Natural History Museum, Chicago.

Martin, Paul S., and John B. Rinaldo
1960 *Excavations in the Upper Little Colorado Drainage, Eastern Arizona*. Fieldiana: Anthropology Volume 51(1). Chicago Natural History Museum, Chicago.
Martin, Paul S., and James Schoenwetter
1960 America's Oldest Cornfield. *Science* 132(3418):33–34.
Matson, R. G.
1991 *The Origins of Southwestern Agriculture*. University of Arizona Press, Tucson.
1999 The Spread of Maize to the Colorado Plateau. *Archaeology Southwest* 13:10–11.
2002 The Spread of Maize Agriculture in the U.S. Southwest. In *Examining the Farming/Language Dispersal Hypothesis,* edited by Peter Bellwood and Colin Renfrew, pp. 341–56. McDonald Institute for Archaeological Research, University of Cambridge, Cambridge, UK.
2006 What Is Basketmaker II? *Kiva* 72:149–65.
2007 The Archaic Origins of the Zuni. In *Zuni Origins: Toward a New Synthesis of Southwestern Archaeology,* edited by D. A. Gregory and D. R. Wilcox, pp. 97–117. University of Arizona Press, Tucson.
Matson, R. G., and Brian Chisholm
1991 Basketmaker II Subsistence: Carbon Isotopes and Other Dietary Indicators from Cedar Mesa, Utah. *American Antiquity* 56:444–59.
Mauldin, Raymond R.
1993 The Relationship between Groundstone and Agricultural Dependence in Western New Mexico. *Kiva* 58:317–44.
1996 Exploring Patterns in Late Archaic and Early Ceramic Period Residential Occupation in the Northern Chihuahuan Desert. In *Early Formative Adaptations in the Southwest,* edited by Barbara J. Roth, pp. 85–97. Monographs in World Prehistory No. 25, Prehistory Press, Madison, WI.
McBrinn, Maxine
2005 *Social Identities among Archaic Mobile Hunters and Gatherers in the American Southwest.* Archaeological Series 197. Arizona State Museum, University of Arizona, Tucson.
McGuire, Kelly R., and William R. Hildebrandt
2005 Re-thinking Great Basin Foragers: Prestige Hunting and Costly Signaling during the Middle Archaic Period. *American Antiquity* 70:695–712.
Meltzer, David J.
1995 Modelling the Prehistoric Response to Altithermal Climates on the Southern High Plains. In *Ancient Peoples and Landscapes*, edited by E. Johnson, pp. 349–68. Museum of Texas Tech University, Lubbock.
Merrill, William L., Robert J. Hard, Jonathan B. Mabry, Gayle J. Fritz, Karen R. Adams, John R. Roney, and A. C. MacWilliams
2009 The Diffusion of Maize to the Southwestern United States and Its Impact. *Proceedings of the National Academy of Sciences* 106(50):21019–26.
Meskell, Lynn
2001 Archaeologies of Identity. In *Archaeological Theory Today*, edited by Ian Hodder, pp. 187–213. Polity Press, Oxford.
2002 The Intersections of Politics and Identity. *Annual Review of Anthropology* 31:279–301.
Miljour, Heather J., and Edgar K. Huber
2005 LA 137258 (Old Corn Site). In *Fence Lake Project. Volume 1: Introduction and Site Descriptions*, edited by Edgar K. Huber and Carla R. Van West, pp. 5.1–5.86. Technical Series No. 84. Statistical Research, Inc., Tucson, AZ.

Miller, Myles R.

In press Archaic Transitions and Transformations in the Jornada Mogollon Region of Southern New Mexico and Western Texas. In *Southwest Archaic*, edited by Bradley J. Vierra. University of Utah Press.

Miller, Myles R., and Nancy A. Kenmotsu

2004 Prehistory of the Jornada Mogollon and Eastern Trans-Pecos Regions of West Texas. In *The Prehistory of Texas*, edited by T. K. Perttula, pp. 205–65. Texas A&M University Press, College Station.

Minnis, Paul E.

1985 Domesticating People and Plants in the Greater Southwest. In *Prehistoric Food Production in North America*, edited by R. Ford, pp. 309–39. Anthropological Papers No. 75. Museum of Anthropology, Ann Arbor, MI.

1992 Earliest Plant Cultivation in the Desert Borderlands of North America. In *The Origins of Agriculture*, edited by C. W. Cowan and P. J. Watson, pp. 121–42. Smithsonian Institution Press, Washington, DC.

2004 People and Plants in Ancient Western North America: An Introduction. In *People and Plants in Ancient Western North America*, edited by Paul E. Minnis, pp. 1–16. University of Arizona Press, Tucson.

Mithen, Steven J.

1990 *Thoughtful Foragers: A Study of Prehistoric Decision Making*. Cambridge University Press, Cambridge, UK.

Moreno, Teresa K.

2000 Accelerator Mass Spectrometry Dates from McEuen Cave. *Kiva* 65:341–60.

Morgan, Christopher

2015 Is It Intensification Yet? Current Archaeological Perspectives on the Evolution of Hunter-gatherer Economies. *Journal of Archaeological Research* 23:163–213.

Morris, Earl H., and Robert F. Burgh

1954 *Basketmaker II Sites Near Durango, Colorado*. Carnegie Institution of Washington Publication 604, Washington, DC.

Mowrer, Kathy

2006 Basketmaker II Mortuary Practices: Social Differentiation and Regional Variation. *Kiva* 72:259–81.

Murrell, Monica L., and Bradley J. Vierra (editors)

2014 *Bridging the Basin: Land Use and Social History in the Southern Chuska Valley*. Technical Report No. 13-69, Statistical Research, Inc., Albuquerque, NM

Nials, Fred L.

2008 Geomorphology and Stratigraphy. In *Las Capas: Early Irrigation and Sedentism in a Southwestern Floodplain*, edited by Jonathan B. Mabry, pp. 35–53. Anthropological Paper No. 28. Center for Desert Archaeology, Tucson, AZ.

2015 Life on the Floodplain: The Promise and Perils of Prehistoric Irrigation Agriculture at Las Capas, AZ AA:12:111 (ASM). In *Implements of Change: Tools, Subsistence, and the Built Environment of Las Capas, an Early Agricultural Irrigation Community in Southern Arizona*, edited by James M. Vint, pp. 419–68. Anthropological Papers No. 51, Archaeology Southwest, Tucson.

O'Connell, James F., and Kristen Hawkes

1981 Alyawara Plant Use and Optimal Foraging Theory. In *Hunter-Gatherer Foraging Strategies: Ethnographic and Archaeological Analysis,* edited by Bruce Winterhalder and E. A. Smith, pp. 99–125. University of Chicago Press, Chicago.

O'Laughlin, Thomas C.

1980 *The Keystone Dam Site and Other Archaic and Formative Sites in Northwest El Paso, Texas.* Publications in Anthropology No. 8. Centennial Museum, University of Texas at El Paso, El Paso.

O'Shea, John

1989 The Role of Wild Resources in Small-Scale Agricultural Systems: Tales from the Lakes and Plains. In *Bad Year Economics,* edited by Paul Halstead and John O'Shea, pp. 57–67. Cambridge University Press, Cambridge, UK.

Parry, William J. and F. E. Smiley

1990 Hunter-Gatherer Archaeology in Northeastern Arizona and Southeastern Utah. In *Perspectives on Southwestern Prehistory,* edited by Paul E. Minnis and Charles L. Redman, pp. 47–56. Westview Press, Boulder, CO.

Parry, William J., F. E. Smiley, and G. R. Burgett

1994 The Archaic Occupation of Black Mesa, Arizona. In *Archaic Hunter-Gatherer Archaeology in the American Southwest,* edited by Bradley J. Vierra, pp. 185–230. Contributions in Anthropology No. 13(1). Eastern New Mexico University, Portales, NM.

Phillips, Bruce G., Gregory E. Berg, Lourdes Aguila, and Barbara S. Macnider

2001 *Data Recovery at AZ U:5:33(ASM) within the Pima Freeway Corridor, Phoenix, Maricopa County, Arizona.* Archaeological Consulting Services, Tempe, AZ.

Phillips, David A., Jr.

2009 Adoption and Intensification of Agriculture in the North American Southwest: Notes Toward a Quantitative Approach. *American Antiquity* 74:691–707.

Phillips, David A., Mark C. Slaughter, and Susan Bierer

1993 *Archaeological Studies at Kartchner Caverns State Park, Cochise County, Arizona.* SWCA Archaeological Report 93-26, Tucson, AZ.

Piperno, Dolores R., and Kent V. Flannery

2001 The Earliest Archaeological Maize (*Zea mays L.*) from Highland Mexico: New Accelerator Mass Spectrometry Dates and Their Implications. *Proceedings of the National Academy of Science* 98(4):2101–3.

Piperno, Dolores R., Anthrony J. Ranere, Irene Holst, Jose Iriarte, and Ruth Dickau

2009 Starch Grain and Phytolith Evidence for Early Ninth Millenium B.P. Maize from the Central Balsas River Valley, Mexico. *Proceedings of the National Academy of Sciences* 106:5019–24.

Porter, Claire C., and Frank W. Marlowe

2007 How Marginal Are Forager Habitats? *Journal of Archaeological Science* 34:59–68.

Post, Stephen S.

2002 Emerging from the Shadows: The Archaic Period in the Northern Rio Grande. In *Traditions, Transitions, and Technologies: Themes in Southwestern Archaeology,* edited by Sarah H. Schlanger, pp. 33–46. University Press of Colorado, Boulder.

2013 Transitional Archaic and Emergent Agricultural Settlement in Lowland-Upland Settings of the Northern Rio Grande, New Mexico. In *From Mountain Top to Valley Bottom, Understanding Past Land Use in the Northern Rio Grande Valley, New Mexico,* edited by Bradley J. Vierra, pp. 80–94. University of Utah Press, Salt Lake City.

Potter, James M.

2010a Paleoindian and Archaic Sites. In *Animas-La Plata Project: Final Synthetic Report, Volume XVI.* SWCA Anthropological Research Papers 10, pp. 35–49. SWCA Environmental Consultants, Phoenix, AZ.

2010b Basketmaker II Sites. In *Animas-La Plata Project: Final Synthetic Report, Volume XVI*, edited by James M. Potter, pp. 51–75. SWCA Anthropological Research Papers 10. SWCA Environmental Consultants, Phoenix, AZ.

Powell, Shirley, and Francis E. Smiley

2002 *Prehistoric Culture Change on the Colorado Plateau: Ten Thousand Years on Black Mesa*. University of Arizona Press, Tucson.

Pryor, Frederic L.

1986 The Adoption of Agriculture: Some Theoretical and Empirical Evidence. *American Anthropologist* 88:879–95.

Railey, Jim A.

2016 Early Farming and the Fate of Archaic Hunter-gatherers in the Albuquerque Basin. In *Late Holocene Foragers and Farmers in the Desert West*, edited by Barbara J. Roth and Maxine McBrinn, pp. 16–57. University of Utah Press, Salt Lake City.

Ranere, Anthony J., Dolores R. Piperno, Irene Holst, Ruth Dickau, and Jose Iriarte

2009 The Cultural and Chronological Context of Early Holocene Maize and Squash Domestication in the Central Balsas River Valley, Mexico. *Proceedings of the National Academy of Sciences* 106(13):5014–18.

Redding, Richard W.

1988 A General Explanation of Subsistence Change: From Hunting and Gathering to Food Production. *Journal of Anthropological Archaeology* 7:56–97.

Reed, Paul F. (editor)

2000 *Foundations of Anasazi Culture, The Basketmaker-Pueblo Transition*. University of Utah Press, Salt Lake City.

Roberts, Heidi R.

n.d. The Jackson Flat Reservoir Project: Early Maize Dates, Antechambers, Floor Vaults, and an Oversized Pithouse at a Virgin Branch Settlement in Kanab, Utah. In *Interaction and Connectivity in the Greater Southwest*, edited by Karen G. Harry and Barbara J. Roth. University Press of Colorado, under review.

Rocek, Thomas R.

1995 Sedentarization and Agricultural Dependence: Perspectives from the Pithouse-to-Pueblo Transition in the American Southwest. *American Antiquity* 60:218–39.

1996 Sedentism and Mobility in the Southwest. In *Interpreting Southwestern Diversity: Underlying Principles and Overarching Patterns*, edited by P. R. Fish and J. J. Reid, pp. 17–22. Arizona State University Anthropological Research Papers No. 48, Tempe, AZ.

Rogers, Malcolm J.

1939 *Early Lithic Industries of the Lower Basin of the Colorado River and Adjacent Desert Areas*. San Diego Museum Papers No. 3. Museum of Man, San Diego.

Roney, John R., and Robert J. Hard

2002 Early Agriculture in Northwestern Chihuahua. In *Traditions, Transitions, and Technologies: Themes in Southwestern Archaeology*, edited by Sarah H. Schlanger, pp. 163–80. University of Press of Colorado, Boulder.

Roney, John R., Robert J. Hard, A. C. MacWilliams, and Mary Whisenhunt

2014 Initial Investigations at the Round Mountain Cerro de Trincheras Site in Southeastern Arizona. Paper presented at the 18th Biennial Mogollon Archaeology Conference, Las Cruces, New Mexico.

Roth, Barbara J.

1992 Sedentary Agriculturalists or Mobile Hunter-Gatherers? Recent Evidence on the Late Archaic Occupation of the Northern Tucson Basin. *Kiva* 57:291–314.

1995 Late Archaic Occupation of the Upper Bajada: Excavations at AZ AA: 12:84 (ASM). *Kiva* 67:59–79.

1996 Regional Land Use in the Late Archaic of the Tucson Basin. In *Early Formative Adaptations in the Southern Southwest,* edited by Barbara J. Roth, pp. 37–48. Monographs in World Archaeology No. 25. Prehistory Press, Madison, WI.

2006 The Role of Gender in the Adoption of Agriculture in the Southern Southwest. *Journal of Anthropological Research* 62:513–38.

2014 Foragers, Farmers, and In Between: Variability in the Late Archaic of the Southern Southwest. In *Archaeology in the Great Basin and Southwest: Papers in Honor of Don D. Fowler,* edited by Nancy J. Parezo and Joel C. Janetski, pp. 98–108. University of Utah Press, Salt Lake City.

2016 Were They Sedentary and Does It Matter? Early Farmers in the Tucson Basin. In *Late Holocene Foragers and Farmers in the Desert West,* edited by Barbara J. Roth and Maxine McBrinn, pp. 108–35. University of Utah Press, Salt Lake City.

Roth, Barbara J., and Justin DeMaio

2014 Early and Middle Archaic Foragers in the Sonoran Desert: Investigations at the Ruelas Ridge Site. *Kiva* 80:115–36.

Roth, Barbara J., and Andrea Freeman

2008 The Middle Archaic Period and the Transition to Agriculture in the Sonoran Desert of Southern Arizona. *Kiva* 73:321–54.

Roth, Barbara J., and Bruce B. Huckell

1992 Cortaro Points and the Archaic of Southern Arizona. *Kiva* 57:353–70.

Roth, Barbara J., and Kevin Wellman

2001 New Insights into the Early Agricultural Period in the Tucson Basin: Excavations at the Valley Farms Site. *Kiva* 67:59–79.

Rouse, Irving

1958 The Inference of Migrations from Anthropological Evidence. Migrations in New World Culture History. *University of Arizona Social Science Bulletin* 27:63–68.

Rowley-Conwy, Peter, and Robert H. Layton

2011 Foraging and Farming as Niche Construction: Stable and Unstable Adaptations. *Philosophical Transactions of the Royal Society of London Series B* 366(1566):849–62.

Sackett, James R.

1977 The Meaning of Style in Archaeology: A General Model. *American Antiquity* 42:369–80.

Sayles, E.B.

1945 *The San Simon Branch, Excavations at Cave Creek and in the San Simon Valley.* Medallion Papers No. 34, Gila Pueblo, Globe, AZ.

1983 *The Cochise Cultural Sequence in Southeastern Arizona.* Anthropological Papers No. 42. University of Arizona, Tucson.

Sayles, E. B., and Ernst Antevs

1941 *The Cochise Culture.* Medallion Papers No. 29, Gila Pueblo, Globe, AZ.

Schaafsma, Polly

1980 *Indian Rock Art of the Southwest.* School of American Research, Santa Fe, NM.

Schmader, Matthew

2001 *Gimme Shelter: Uncovering Archaic Structures in Rio Rancho and Santa Fe, New Mexico.* Rio Grande Consultants, Albuquerque, NM.

Schmidt, Kari M.
2008 *An Assessment of Settlement and Subsistence in Emergent Agricultural Economies in the Tucson Basin, United States, and Chihuahua, Mexico.* Unpublished Ph.D. dissertation, Department of Anthropology, University of New Mexico, Albuquerque.

Schroedl, Alan R.
1976 The Archaic of the Northern Colorado Plateau. Unpublished Ph.D. dissertation, Department of Anthropology, University of Utah, Salt Lake City.

Schroth, Adella B.
1994 *The Pinto Point Controversy in the Western United States.* Unpublished Ph.D. dissertation, Department of Anthropology, University of California, Riverside.

Seymour, Deni J., Jeffrey H. Hokanson, and Vicky J. T. Cunningham
1997 *Excavations at Lru-Kachreu and Other Sites at the Sandoval County Landfill.* Lone Mountain Archaeological Services, Albuquerque, NM.

Shackley, M. Steven
1986 Lithic Technology and Mobility Strategies at Picacho. In *Prehistoric Hunter-Gatherers of South Central Arizona: The Picacho Reservoir Archaic Project,* edited by Frank Bayham, Donald H. Morris, and M. Steven Shackley, pp. 109–56. Anthropological Field Studies No. 13. Department of Anthropology, Office of Cultural Resource Management, Arizona State University, Tempe.

Shipek, Florence C.
1989 An Example of Intensive Plant Husbandry: The Kumeyaay of Southern California. In *Foraging and Farming: The Evolution of Plant Exploitation,* edited by D. R. Harris and G. C. Hillman, pp. 159–70. Unwin Hyman Press, London.

Simmons, Alan H.
1982 *Prehistoric Adaptive Strategies in the Chaco Region, Northwestern New Mexico,* assembled by Alan H. Simmons. Navajo Nation Papers in Anthropology No. 9. Window Rock, AZ.
1986 New Evidence for the Early Use of Cultigens in the American Southwest. *American Antiquity* 51:73–89.
2006 Early People, Early Maize, and Late Archaic Ecology in the Southwest. In *Environmental Change and Human Adaptation in the Ancient American Southwest,* edited by David Doyel and Jeffrey Dean, pp. 10–25. University of Utah Press, Salt Lake City.

Simms, Steven R.
1985 Acquisition Cost and Nutrition Data on Great Basin Resources. *Journal of California and Great Basin Anthropology* 7:117–25.

Sliva, R. Jane
2005 Developments in Flaked Stone Technology during the Transition to Agriculture. In *Material Cultures and Lifeways of Early Agricultural Communities in Southern Arizona,* edited by R. Jane Sliva, pp. 47–98. Anthropological Papers No. 35. Center for Desert Archaeology, Tucson, AZ.
2015 *Projectile Points of the Early Agricultural Southwest.* Archaeology Southwest, Tucson.

Smalley, John, and Michael Blake
2003 Sweet Beginnings: Stalk Sugar and the Domestication of Maize. *Current Anthropology* 44:675–703.

Smiley, Francis E.
1994 The Agricultural Transition in the Northern Southwest: Patterns in the Current Chronometric Data. *Kiva* 60:165–89. University of Arizona Press, Tucson.

2002 Black Mesa before Agriculture: Paleoindian and Archaic Evidence. In *Prehistoric Culture Change on the Colorado Plateau: Ten Thousand Years on Black Mesa,* edited by Shirley Powell and Francis E. Smiley, pp. 15–34. University of Arizona Press, Tucson.

Smith, Bruce

1995 *The Emergence of Agriculture.* Freeman, New York.

1997 The Initial Domestication of *Cucurbita pepo* in the Americas 10,000 Years Ago. *Science* 276(5314):932–34.

2001a Documenting Plant Domestication: The Consilience of Biological and Archaeological Approaches. *Proceedings of the National Academy of Sciences* 98: 1324–26.

2001b Low-Level Food Production. *Journal of Archaeological Research* 9(1):1–43.

2005 Reassessing Coxcatlan Cave and the Early History of Domesticated Plants in Mesoamerica. *Proceedings of the National Academy of Sciences* 102:9438–45.

Smith, Craig S.

2003 Hunter-gatherer Mobility, Storage, and Houses in a Marginal Environment: An Example from the Mid-Holocene of Wyoming. *Journal of Anthropological Archaeology* 22:162–89.

Smith, Craig S., and Lance M. McNees

1999 Facilities and Hunter-Gatherer Long-Term Land Use Patterns: An Example from Southwest Wyoming. *American Antiquity* 64:117–36.

Speller, Camilla F., Brian M. Kemp, Scott D. Wyatt, Cara Monroe, William D. Lipe, Ursula M. Arndt, and Dongya Y. Yang.

2010 Ancient Mitochondrial DNA Analysis Reveals Complexity of Indigenous North American Turkey Domestication. *Proceedings of the National Academy of Sciences* 107(7):2807–12.

Speth, John, and Katherine A. Spielmann

1983 Energy Source, Protein Metabolism, and Hunter-Gatherer Subsistence Strategies. *Journal of Anthropological Archaeology* 2:1–31.

Stevens, Michelle N., and R. Jane Sliva

2002 Empire Points: An Addition to the San Pedro Phase Lithic Assemblage. *Kiva* 67: 297–326.

Steward, Julian

1933 Ethnography of the Owens Valley Paiute. *University of California Publications in American Ethnography and Ethnology* 33:233–50.

1938 *Basin-Plateau Aboriginal Sociopolitical Groups.* Bureau of American Ethnology Bulletin 120. Smithsonian Institution, Washington, DC.

Stone, Tammy

2003 Social Identity and Ethnic Interaction in the Western Pueblos of the American Southwest. *Journal of Archaeological Method and Theory* 10:31–67.

Stutz, Aaron J.

2012 Culture and Politics, Behavior and Biology: Seeking Synthesis among Fragmentary Anthropological Perspectives on Hunter-Gatherers. *Reviews in Anthropology* 41(1):23–69.

Sutton, Mark Q.

1984 The Productivity of *Pinus monophylla* and Modeling Great Basin Subsistence Strategies. *Journal of California and Great Basin Anthropology* 6:240–45.

Tagg, Martyn D.

1996 Early Cultigens from Fresnal Shelter, Southeastern New Mexico. *American Antiquity* 61:311–24.

Thiel, J. Homer, and Jonathan B. Mabry (editors)

2006a *Rio Nuevo Archaeology, 2000–2003: Investigations at the San Agustin Mission and Mission Gardens, Tucson Presidio, Tucson Pressed Brick Company, and Clearwater Site.* Technical Report No. 2004-11. Desert Archaeology, Inc., Tucson, AZ.

Thiel, J. Homer, and Jonathan B. Mabry

2006b Summary: The Archaeology of a Changing Community. In *Rio Nuevo Archaeology, 2000–2003: Investigations at the San Agustin Mission and Mission Gardens, Tucson Presidio, Tucson Pressed Brick Company, and Clearwater Site,* edited by J. Homer Thiel and Jonathan B. Mabry, pp. 22.1–22.13. Technical Report No. 2004-11. Desert Archaeology, Inc., Tucson, AZ.

Thomas, David H.

1972 Western Shoshone Ecology: Settlement Patterns and Beyond. In *Great Basin Cultural Ecology: A Symposium,* edited by Donald D. Fowler, pp. 135–53. University of Nevada Press, Reno.

Thurtle, Mary Charlotte, Andrea Kayser, and Kevin D. Wellman

2008 Summary of Excavation Results. In *The Valley Farms Site, Prehistoric Floodplain Agriculture on the Santa Cruz River in Southern Arizona,* edited by Kevin D. Wellman, pp. 17–58. SWCA Anthropological Research Paper No. 11, Tucson.

Toll, Mollie S., and Anne C. Cully

1983 Archaic Subsistence in the Four Corners Area: Evidence for a Hypothetical Seasonal Round. In *Economy and Interaction along the Lower Chaco River,* edited by Patrick Hogan and Joseph C. Winter, pp. 385–92. Office of Contract Archaeology, University of New Mexico, Albuquerque, NM.

1994 Archaic Subsistence and Seasonal Population Flow in Northwest New Mexico. In *Archaic Hunter-Gatherer Archaeology in the American Southwest,* edited by Bradley J. Vierra, pp. 103–20. Eastern New Mexico Contributions to Anthropology 13(1). Portales, NM.

Tringham, Ruth E.

1991 Households with Faces: The Challenge of Gender in Prehistoric Architectural Remains. In *Engendering Archaeology: Women and Prehistory,* edited by Joan Gero and Margaret W. Conkey, pp. 93–131. Blackwell Press, Oxford.

Turnbow, Christopher A.

2000 *A Highway Through Time: Archaeological Investigations along NM 90, in Grant and Hidalgo Counties, New Mexico.* U.S. Forest Service Report 03-06-98-005b. TRC, Albuquerque, NM.

Turnbow, Christopher, Alexander Kurota, and Richard C. Chapman

2008 *Preliminary Report on Archeological Data Recovery at LA 85777, LA 85776, and LA 85779between Access Roads 2 and 3, Luna County, New Mexico.* OCA/UNM Report No. 185-986A.Office of Contract Archeology, University of New Mexico, Albuquerque

Turner, Raymond M., and David E. Brown

1982 Sonoran Desertscrub. In *Biotic Communities of the American Southwest-United States and Mexico.* Desert Plants 4(1-4):181–221.

Upham, Steadman, and Richard S. MacNeish

1993 The Evolution of Maize in the Jornada Region of New Mexico and Its Implications of the Southwest. In *Preliminary Investigations of the Archaic in the Region of Las Cruces, New Mexico,* edited by Richard S. MacNeish, pp. 105–16. U. S. Army Air Defense Artillery Center, Fort Bliss, TX.

Upham, Steadman, Richard S. MacNeish, Walton C. Galinat, and Christopher M. Stevenson

1987 Evidence Concerning the Origin of Maiz de Ocho. *American Anthropologist* 89:410–19.

Van Devender, Thomas R.

1990 Late Quaternary Vegetation and Climate of the Sonoran Desert, United States and Mexico. In *Packrat Middens: The Last 40,000 Years of Biotic Change,* edited by Julio L. Betancourt, Thomas R. Van Devender, and Paul S. Martin, pp. 134–65. University of Arizona Press, Tucson.

Varien, Mark D.

1999 *Sedentism and Mobility in a Social Landscape: Mesa Verde and Beyond.* University of Arizona Press, Tucson.

Verhoeven, Marc

2002 Transformations of Society: The Changing Role of Ritual and Symbolism in the PPNB and the PN in the Levant, Syria, and South-east Anatolia. *Paléorient* 28:5–13.

Vierra, Bradley

2005 Late Archaic Stone Tool Technology across the Borderlands. In *The Late Archaic Across the Borderlands: Foraging to Farming,* edited by Bradley J. Vierra, pp. 187–218. University of Texas Press, Austin.

2008 Early Agriculture on the Southwestern Periphery of the Colorado Plateau: Diversity in Tactics. In *Archaeology Without Borders: Contact, Commerce, and Change in the U.S. Southwest and Northwestern Mexico,* edited by Laurie D. Webster and Maxine E. McBrinn, pp. 71–88. University Press of Colorado, Boulder.

2009 *Keystone in Context: A Significant Archaic Period Site in El Paso, Texas.* Keystone Heritage Park, El Paso, TX.

2013 *From Mountaintop to Valley Bottom: Understanding Past Land-Use in the Northern Rio Grande, New Mexico.* University of Utah Press, Salt Lake City.

Vierra, Bradley and Maxine McBrinn

2016 Resistant Foragers: Foraging and Maize Cultivation in the Northern Rio Grande Valley. In *Late Holocene Foragers and Farmers in the Desert West,* edited by Barbara J. Roth and Maxine E. McBrinn, pp. 58–77. University of Utah Press, Salt Lake City.

Vierra, Bradley J., Margaret A. Jodry, M. Steven Shackley, and Michael J. Dilley

2012 Late Paleoindian and Early Archaic Foragers in the Northern Southwest. In *From the Pleistocene to the Holocene: Human Organization and Cultural Transformations in Prehistoric North America,* edited by C. Britt Bousman and Bradley J. Vierra, pp. 171–95. Texas A&M University Press, College Station, TX.

Vierra, Bradley J., and Richard I. Ford

2006 Early Maize Agriculture in the Northern Rio Grande Valley, New Mexico. In *Histories of Maize: Multidisciplinary Approaches to the Prehistory, Domestication, and Evolution of Maize,* edited by J. Staller, pp. 497–510. Elsevier Press, New York.

2007 Foragers to Farmers in the Northern Rio Grande, New Mexico. *Kiva* 73:117–30.

Vierra, Bradley and Teralene Foxx

2010 Archaic Upland Resource Use: The View from the Pajarito Plateau. In *Between Mountains—Beyond the Mountains: Papers in Honor of Paul Williams,* pp. 153–66. Papers of the Archaeological Society of New Mexico, Vol. 35. Albuquerque, NM.

Vint, James M. (editor)
2015 *Implements of Change: Tools, Subsistence, and the Built Environment of Las Capas, an Early Agricultural Irrigation Community in Southern Arizona.* Anthropological Papers No. 51, Archaeology Southwest, Tucson.

Vint, James M., and Fred L. Nials
2015 *The Anthropogenic Landscape of Las Capas, an Early Agricultural Irrigation Community in Southern Arizona.* Anthropological Papers No. 50, Tucson.

Warren, Claude N.
1980 Pinto Points and Problems in Mojave Desert Archaeology. In *Anthropological Papers in Memory of Earl H. Swanson Jr.*, edited by L. B. Harten, C. N. Warren, and D. R. Tuohy, pp. 67–76. Idaho Museum of Natural History, Pocatello.

Waters, Michael R.
1986 *The Geoarchaeology of Whitewater Draw, Arizona.* Anthropological Papers No. 45. University of Arizona Press, Tucson.

Waters, Michael R., and Anne I. Woosley
1990 The Geoarchaeology and Preceramic Prehistory of the Willcox Basin, SE Arizona. *Journal of Field Archaeology* 17:163–75.

Watson, James T.
2007 *Cavities on the Cob: Dental Health and the Agricultural Transition in Sonora, Mexico.* Unpublished Ph.D. dissertation, Department of Anthropology, UNLV, Las Vegas, NV.

Watson, James T., and Rachael M. Byrd
2015 A Bioarchaeological Perspective on Change and Continuity in an Early Agricultural Community. In *Implements of Change: Tools, Subsistence, and the Built Environment of Las Capas, an Early Agricultural Irrigation Community in Southern Arizona*, edited by James M. Vint, pp. 377–88. Anthropological Papers No. 51, Archaeology Southwest, Tucson.

Webster, Laurie D., and Kelley A. Hays-Gilpin
1994 New Trails for Old Shoes: Sandals, Textiles, and Baskets in Basketmaker Culture. *Kiva* 60:313–27.

Wegener, Robert M., and Richard Ciolek-Torello
2011 *Excavations of Prehistoric Sites in the Queen Valley to Queen Creek Area.* The U.S. 60 Archaeological Project: Early Agricultural, Formative, and Historical-Period Use of the Upper Queen Creek Region, Vol. 2. Technical Series 92. Statistical Research, Tucson, AZ.

Wegener, Robert M., and John D. Hall (editors)
2014 *Analysis and Interpretations.* In The Luke Air Force Base Solar-Power Array Archaeological Project: 5000 Years of Aboriginal Land Use in the Western Phoenix Basin, Vol. 1. Technical Report 14-18, Statistical Research, Inc., Tucson, AZ.

Welch, John R.
1991 From Horticulture to Agriculture in the Late Prehistory of the Grasshopper Region, Arizona. In *Mogollon V*, edited by Patrick H. Beckett, pp. 75–92. COAS Publishing, Las Cruces, NM.

Wellman, Kevin D. (editor)
2008 *The Valley Farms Site, Prehistoric Floodplain Agriculture on the Santa Cruz River in Southern Arizona.* SWCA Anthropological Research Paper No. 11, Tucson.

Wendorf, Fred, and T. Thomas
1951 Early Man Sites near Concho, Arizona. *American Antiquity* 17:107–14.

Whalen, Michael E.

1994 Moving Out of the Archaic on the Edge of the Southwest. *American Antiquity* 59:622–38.

Whalen, Norman M.

1971 *Cochise Culture Sites in the Central San Pedro Drainage, Arizona.* Unpublished Ph.D. Dissertation, University of Arizona, Tucson.

1973 Agriculture and the Cochise. *Kiva* 39:89–95.

Wheaton, Gene

2013 Pitstructures or Brush Huts? An Analysis of 20 Recently Discovered Residential Structures in the Jornada Mogollon Area of Southern New Mexico and West Texas. *Kiva* 79:55–81.

Whittlesey, Stephanie M.

2010 Sedentism, Settlement Pattern, and Seasonality: Thoughts on Recurrent Sedentism and the Making of Place. In *Recurrent Sedentism and the Making of Place: Archaeological Investigations at Las Capas, A Preceramic Farming Community in the Tucson Basin, Southern Arizona,* edited by Stephanie M. Whittlesey, S. Jerome Hesse, and Michael S. Foster, pp. 509–20. Cultural Resources Report No. 07-556. SWCA Environmental Consultants, Inc., Tucson.

Whittlesey, Stephanie M., and Richard Ciolek-Torrello

1996 The Archaic–Formative Transition in the Tucson Basin. In *Early Formative Adaptations in the Southern Southwest,* edited by Barbara J. Roth, pp. 49–64. Monographs in World Archaeology No. 25. Prehistory Press, Madison, WI.

Whittlesey, Stephanie M., and Michael S. Foster

2010 The Middle Preceramic Period at Las Capas. In *Recurrent Sedentism and the Making of Place: Archaeological Investigations at Las Capas, a Preceramic Period Farming Community in the Tucson Basin, Southern Arizona,* edited by Stephanie M. Whittlesey, S. Jerome Hesse, and Michael S. Foster, pp. 497–508. Cultural Resource Report No. 07-556. SWCA Environmental Consultants, Tucson, AZ.

Whittlesey, Stephanie M., S. Jerome Hesse, and Michael S. Foster (editors)

2010 *Recurrent Sedentism and the Making of Place: Archaeological Investigations at Las Capas, a Preceramic Period Farming Community in the Tucson Basin, Southern Arizona.* Cultural Resource Report No. 07-556. SWCA Environmental Consultants, Tucson, AZ.

Wiley, Gordon R., and Phillip Phillips

1958 *Method and Theory in American Archaeology.* University of Chicago Press.

Wills, Wirt H.

1988a *Early Prehistoric Agriculture in the American Southwest.* School of American Research Press, Santa Fe, NM.

1988b Early Agriculture and Sedentism in the American Southwest: Evidence and Interpretations. *Journal of World Prehistory* 2:445–88.

1989 Patterns of Prehistoric Food Production in West-Central New Mexico. *Journal of Anthropological Research* 45:139–57.

1992 Plant Cultivation and the Evolution of Risk-Prone Economies in the Prehistoric American Southwest. In *Transitions to Agriculture in Prehistory,* edited by Anne B. Gebauer and T. Douglas Price, pp. 153–76. Monographs in World Archaeology. Prehistory Press, Madison, WI.

1995 Archaic Foraging at the Beginning of Food Production in the American Southwest. In *Last Hunters-First Farmers: New Perspectives on the Prehistoric Transition to Agriculture,* edited by T. Douglas Price and Anne B. Gebauer, pp. 215–346. School of American Research Press, Santa Fe, NM.

2006 Review of *The Late Archaic Across the Borderlands: From Foraging to Farming,* by Bradley Vierra. *Kiva* 72:119–27.

Wills, Wirt H., and Bruce B. Huckell

1994 Economic Implications of Changing Land-Use Patterns in the Late Archaic. In *Themes in Southwest Prehistory,* edited by George J. Gumerman, pp. 3–50. Advanced Seminar Series, School of American Research Press, Santa Fe, NM.

Winter, Joseph C., and Patrick F. Hogan

1986 Plant Husbandry in the Great Basin and Adjacent Northern Colorado Plateau. In *Anthropology of the Desert West: Essays in Honor of Jesse D. Jennings,* edited by Carol J. Condie and Don D. Fowler, pp. 117–44. Anthropological Papers No. 110. University of Utah Press, Salt Lake City.

Winterhalder, Bruce, and Carol Goland

1997 An Evolutionary Ecology Perspective on Diet Choice, Risk, and Plant Domestication. In *People, Plants, and Landscapes: Studies in Paleoethnobotany,* edited by Kristin J. Gremillion, pp. 123–60. University of Alabama Press, Tuscaloosa, AL.

Wöcherl, Helga

2005 Pits and the Use of Extramural Space in Early Farming Communities. In *Material Cultures and Lifeways of Early Agricultural Communities in Southern Arizona,* edited by R. Jane Sliva, pp. 19–46. Anthropological Papers No. 35. Center for Desert Archaeology, Tucson, AZ.

Woodson, M. Kyle

1999 Migrations in Late Anasazi Prehistory: The Evidence from the Goat Hill Site. *Kiva* 65:63–84.

Zeder, Melinda (editor)

2006 *Documenting Domestication: New Genetic and Archaeological Paradigms.* University of California Press, Oakland.

Index

Note: Page references for figures and tables are italicized.

About the Author

Barbara J. Roth is professor and chair of the Department of Anthropology at the University of Nevada, Las Vegas. She received her PhD in anthropology from the University of Arizona and began her career studying hunter-gatherers and the transition to farming in the Sonoran Desert. She has worked at Archaic sites throughout southern Arizona and in the Mojave Desert. More recently she has been doing fieldwork in the Mimbres region of southwestern New Mexico to explore what happened to groups after they adopted and became more focused on agriculture.